Biographies of Scientists

THE MAGILL BIBLIOGRAPHIES

The American Presidents, by Norman S. Cohen, 1989
Black American Women Novelists, by Craig Werner, 1989
Classical Greek and Roman Drama, by Robert J. Forman, 1989
Contemporary Latin American Fiction, by Keith H. Brower, 1989
Masters of Mystery and Detective Fiction, by J. Randolph Cox, 1989
Nineteenth Century American Poetry, by Philip K. Jason, 1989
Restoration Drama, by Thomas J. Taylor, 1989
Twentieth Century European Short Story, by Charles E. May, 1989
The Victorian Novel, by Laurence W. Mazzeno, 1989
Women's Issues, by Laura Stempel Mumford, 1989
America in Space, by Russell R. Tobias, 1991
The American Constitution, by Robert J. Janosik, 1991
The Classical Epic, by Thomas J. Sienkewicz, 1991
English Romantic Poetry, by Bryan Aubrey, 1991
Ethics, by John K. Roth, 1991
The Immigrant Experience, by Paul D. Mageli, 1991
The Modern American Novel, by Steven G. Kellman, 1991
Native Americans, by Frederick E. Hoxie and Harvey Markowitz, 1991
American Drama: 1918-1960, by R. Baird Shuman, 1992
American Ethnic Literatures, by David R. Peck, 1992
American Theatre History, by Thomas J. Taylor, 1992
The Atomic Bomb, by Hans G. Graetzer and Larry M. Browning, 1992
Biography, by Carl Rollyson, 1992
The History of Science, by Gordon L. Miller, 1992
The Origin and Evolution of Life on Earth, by David W. Hollar, 1992
Pan-Africanism, by Michael W. Williams, 1992
Resources for Writers, by R. Baird Shuman, 1992
Shakespeare, by Joseph Rosenblum, 1992
The Vietnam War in Literature, by Philip K. Jason, 1992
Contemporary Southern Women Fiction Writers, by Rosemary M. Canfield Reisman and Christopher J. Canfield, 1994
Cycles in Humans and Nature, by John T. Burns, 1994
Environmental Studies, by Diane M. Fortner, 1994
Poverty in America, by Steven Pressman, 1994
The Short Story in English: Britain and North America, by Dean Baldwin and Gregory L. Morris, 1994

Victorian Poetry, by Laurence W. Mazzeno, 1995
Human Rights in Theory and Practice, by Gregory J. Walters, 1995
Energy, by Joseph R. Rudolph, Jr., 1995
A Bibliographic History of the Book, by Joseph Rosenblum, 1995
The Search for Economics as a Science, by the Editors of Salem Press (Lynn Turgeon, Consulting Editor), 1995
Psychology, by the Editors of Salem Press (Susan E. Beers, Consulting Editor), 1995
World Mythology, by Thomas J. Sienkewicz, 1996
Art, Truth, and High Politics: A Bibliographic Study of the Official Lives of Queen Victoria's Ministers in Cabinet, 1843-1969, by John Powell, 1996
Popular Physics and Astronomy, by Roger Smith, 1996
Paradise Lost, by P. J. Klemp, 1996
Social Movement Theory and Research, by Roberta Garner and John Tenuto, 1996
Propaganda in Twentieth Century War and Politics, by Robert Cole, 1996
The Kings of Medieval England, c. 560-1485, by Larry W. Usilton, 1996
The British Novel 1680-1832, by Laurence W. Mazzeno, 1997
The Impact of Napoleon, 1800-1815, by Leigh Ann Whaley, 1997
Cosmic Influences on Humans, Animals, and Plants, by John T. Burns, 1997
One Hundred Years of American Women Writing, 1848-1948, by Jane Missner Barstow, 1997
Vietnam Studies, by Carl Singleton, 1997
British Women Writers, 1700-1850, by Barbara J. Horwitz, 1997
The United States and Latin America, by John A. Britton, 1997
Reinterpreting Russia, by Steve D. Boilard, 1997
Theories of Myth, by Thomas J. Sienkewicz, 1997
Women and Health, by Frances R. Belmonte, 1997
Contemporary Southern Men Fiction Writers, by Rosemary M. Canfield Reisman and Suzanne Booker-Canfield, 1998
Black/White Relations in American History, by Leslie V. Tischauser, 1998
The Creation/Evolution Controversy, by James L. Hayward, 1998
The Beat Generation, by William Lawlor, 1998
Biographies of Scientists, by Roger Smith, 1998
Introducing Canada, by Brian Gobbett and Robert Irwin, 1998
Four British Women Novelists: Anita Brookner, Margaret Drabble, Iris Murdoch, Barbara Pym, by George Soule, 1998

Biographies of Scientists

An Annotated Bibliography

Roger Smith

Magill Bibliographies

The Scarecrow Press, Inc.
Lanham, Md., & London
and
Salem Press
Pasadena, Calif., & Englewood Cliffs, N.J.
1998

SCARECROW PRESS, INC.

Published in the United States of America
by Scarecrow Press, Inc.
4720 Boston Way
Lanham, Maryland 20706

4 Pleydell Gardens, Folkestone
Kent CT20 2DN, England

British Library Cataloguing in Publication Information Available

Library of Congress Cataloging-in-Publication Data

Smith, Roger, 1953 Apr. 19–
Biographies of scientists : an annotated bibliography / Roger Smith.
p. cm.
Includes index.
ISBN 0-8108-3384-0 (alk. paper)
1. Science—Bio-bibliography. 2. Scientists—Biography—Bibliography. I. Title.
Z7404.S64 1998
[.Q1411]
016.5092'2—dc21 98-5954
CIP

The paper used in this publication meets the minimum requirements of American National Standard for Information Sciences—Permanence of Paper for Printed Library Materials, ANSI Z39.48–1984.
Manufactured in the United States of America.

For Dorothy Crambert Smith

with love

Contents

Introduction

> For the essential in the being of a man of my type lies precisely in *what* he thinks and *how* he thinks, not in what he does or suffers.
>
> —Albert Einstein, *Autobiographical Notes*

Scientific biography is a relatively new non-fiction genre. It is a genre unto itself for the same reason that all the others are: It shows us matters we need to understand, about ourselves and about the world around us, in a way the other genres cannot quite match. At the same time it entertains us; that is, a good biography or autobiography of a scientist sparks our intellect or arouses our emotions and suddenly we are swept beyond our own experiences. The 736 books in this bibliography, I hope, will help readers who for any reason—from idle curiosity to professional research—want to lay hands upon books and essays that open up to them the fertile, extraordinary minds of those who, in the words of Alfred North Whitehead, "have the genius to be astonished" by nature.

To be sure, biographies of sages and natural philosophers have appeared since classical times, and readers will find biographical works about many of them cited in this book, for the likes of Aristotle, Euclid, Hippocrates, and Ptolemy have had an overwhelming influence, for better or worse, on generations of thinkers. Still, biographies of scientists are really a feature of the late nineteenth century and twentieth century. Before about 1840—when *scientist* first appeared in written English—such biographies were reserved for exceptionally, shockingly original men (always men), Galileo and Newton, for instance, because their innovations made them celebrities, or notorious. The idea of profiling thinkers of the second rank in order to document how they also contributed to the growth of formal knowledge—that came from the modern era, an era that has placed ever more emphasis upon discovery, progress, technology and applied science, natural laws and constants, and the institutions of science.

Public appetite for biographies of scientists has steadfastly kept pace with the expansion of science. Modern scientific celebrities continue to attract biographers in legions, as they did during their own times, people such as Albert Einstein, Louis Pasteur, and Charles Darwin. But since about 1970 autobiographies and biographies have proliferated, not only for those scientists who became household names but also for those whose reputations were known only to colleagues. The higher scientific literacy of readers almost certainly accounts for some of the increased interest. Yet those twin lures, legend and mystery, are surely the main attractions. Legends abound about scientists and scientific projects, which make engrossing, sometimes titillating reading. For example, there is the legendary brilliance of John von Neumann and Julian Schwinger, the legendary unconventionality of Richard Feynman, the legendary perseverance despite neglect of Barbara McClintock, and the legendary group effort of the Manhattan Project. Something akin to fan clubs have even grown from the legends associated with some figures, such as Nikola Tesla. The mystery resides in those scientific developments that strike non-scientists, at least, as ever more remote from common sense and experience. These developments emerge from what historians often call scientific revolutions: the theory of evolution; the special and general theories of relativity; quantum mechanics; nuclear physics; DNA and genetics; cosmology, grand unified theories (GUTs), and theories of everything (TOEs); and plate tectonics, to name the best known. Biographies and autobiographies by the participants discuss the principles and logic of the discovery or theory and that helps part the technical veil of mystery.

Still, as Nobel Prize-winning microbiologist Salvadore Luria points out in his own autobiography, the legends and mysteries are proper topics of science history and popular science, not biography. It is not careers that make scientific biography distinct and moving either. The professional lives of most scientists, passed in laboratories and classrooms, are pretty dull stuff to outsiders. Even those who took part in a famous event, such as Clarence King's Fortieth Parallel Expedition or the detection of quarks at the Stanford Linear Accelerator, led lives that only became exciting to readers for a short period.

What scientific biography and autobiography can convey better than other types of nonfiction is the *adventure* of discovery—how it feels from the viewpoint of people taken up in it, and with all the mistakes, frustrations, worries, elation, and wonder left in as the adventure unfolds. Popular science and science histories usually pass over such personal experience, but a good biography or autobiography makes the investigation live again for readers who otherwise may have never taken part in scientific work. Bad biography or autobiography degenerates into nostalgia, unctuous praise for old colleagues, and commentary, either in praise or complaint, about the ways a field of research has changed in its administration or aims, but only the very worst, and there are few enough of those in the chapters that follow, completely ignore the vital spark of discovery.

Scope and Design of the Bibliography

This collection is not comprehensive. It is, however, a reasonably thorough introduction to scientific biographies. All the major figures are covered, as are many, many lesser known or even obscure scientists. The coverage concerns only the natural and physical sciences. The first chapter, Multidisciplinary Sources, lists essay collections, dictionaries, encyclopedias, and children's book series that offer biographical information on more than one branch of science. The next eight chapters each cite collections, individual biographies and autobiographies, and series for a single branch or closely related branches. The sciences are divided among seven chapters: Astronomy and Cosmology, Chemistry, Earth Sciences, Life Sciences, Mathematics, Medical Sciences, and Physics. Those scientists are included who conducted research, pioneered in using a technique or carrying out a public program, or created or championed theories. The final chapter, Related Fields, has a few scientists, writers, and administrators who, although not researchers, were in some way important to the development, public awareness, or application of a natural science.

This arrangement reduces the number of chapters by placing together sciences that are related but in reality live apart. Thus, Astronomy and Cosmology also has information about astrophysicists. Earth Sciences includes geology, paleontology, geophysics, meteorology, and oceanography. Life Sciences includes biology, marine biology, botany, entomology, ornithology, ecology, ichthyology, and zoology. Medical Sciences includes researchers in human disease, surgery, anatomy, physiology, nursing, and public health. Finding the best chapter to list biographies of some scientists was sometimes tricky, and occasionally placements may seem debatable. Should Louis Pasteur stand among the chemists, among the biologists because he studied plant diseases, or among the medical researchers because he championed the germ theory of disease? Here he shares quarters with the chemists, which is where he thought he belonged. Should Subrahmanyan Chandrasekhar be with the physicists or the astronomers? Here he joins the astronomers, because although he contributed to several physics subspecialties, he is famous for calculating how stars collapse and die. Readers may disagree with some billets, and for good reason, but they should be able to find information on any scientist quickly nevertheless by simply keeping in mind the scientist's best known work. Should that fail, two indexes conclude the book, one listing the names of scientists and biographers in the collection and another listing scientific subjects and institutions.

Although the scope of the collection is international and concerns people from classical times to the late twentieth century, only sources in English, including translations from other language, are cited. The individual biographies, autobiographies, and collections of biographies have all been published in the twentieth century—or in rare cases are twentieth-century reprints—and are commonly available at public or university libraries. Most, in fact, were published in the last quarter of the century.

The Format of Entries

Each citation of a book or set of volumes has two parts. The first part contains the name of the author or editor, title, edition (if important), number of volumes, place of publication, publisher, and date of publication. The second part is the annotation.

The annotation informs readers of the contents, style, intended audience, and execution of the book. Each entry for a book about an individual scientist opens with a word or phrase indicating what kind of book it is. *Biography* and *autobiography* may seem obvious from the book's title to a careful reader, but the bibliography is designed for quick reference, so repetitive or not, these designations appear. *Memoirs* means that a book treats only a segment of a person's career. *Scientific biography* or *scientific autobiography* warns readers either that most of the contents discuss the principles and conduct of science or that substantial sections require at least the competency in the specific branch of science that a college major in that science should have. *Biography-history* or *biography-popular science* means that the biographer uses his subject, or subjects in the case of collections of biographies, as an opportunity to elucidate the history of science, scientific principles, or use of technology. If a book was written for teenagers or children, the approximate age level accompanies the book-type designation: for example, *Biography for readers 12 years and older*. The age level is always somewhat conservative; skillful younger readers should have no problems, but those who do not read much will want to pay attention to the age level.

Next comes the text of the annotation. Mostly, it summarizes contents, remarks about the life and achievements of the subject, and evaluates the effectiveness of the writing. If the intended audience of the book is not obvious from the title, and such is frequently the case, the annotation indicates that audience, such as historians and other scientists. Readers who want to avoid technical discussions, scholarly documentation, or the formal style of professional disquisition are thereby forewarned, even if they could otherwise understand the contents without special training. To place scientists historically, their birth and death dates appear by their names at least once.

The annotation ends with a list of non-textual contents: whether a book contains photographs, drawings, diagrams, bibliography, glossary, end notes, footnotes, or indexes. For scholars this information can be helpful, because bibliographies and notes point to yet further sources of information. Readers who are neither scholars nor scientists may like to know that a book has a glossary and diagrams to help acquaint them with the technical matters that the main text discusses. At the very end of the annotation is the number of pages of the book; if the citation concerns a multivolume set, the number of volumes has already been noted in the first part of the annotation, and the page numbers are not stated.

The annotations, I trust, will not just prove a useful starting point for research about scientists or science history but also will direct readers to the pleasures of a distinctly modern literature.

WEBSITES

Readers interested in scientists who have recently won the attention of colleagues or the public may not find information about them in books, anthologies, or encyclopedias. If that is the case, readers can try *Scientific American*, which publishes a profile of a scientist in each monthly issue. Or readers might search the Internet: Here are informative websites to start with.

Eric's Treasure Trove of Scientific Biography
http://www.astro.virginia.edu/~eww6n/bios/bios.html
AIP Center for History of Physics
http://ww.aip.org/history/index.html
University of Wisconsin History of Science
http://www.physics.wisc.edu/~shaliz/hyper-weird/history of science.html
University of Melbourne, Australia, History of Science, Technology and Medicine
http://www.asap.univelb.edu.au/hstm/hstm_ove.htm
History of Science Society
http://weber.u.washington.edu/~hssexec/index.html

Institut für Geschicht der Naturwissenschaften (Institute for the History of Science)
http://www.rs.uni-frankfurt.de/~linhard/igne.html
Mathematics of the Seventeenth and Eighteenth Centuries
http://www.maths.tcd.ie/pub/HistMath/People/RBallHist.html
Biographies of Women Mathematicians
http://www.scotlan.edu./lriddle/women/WOMEN.HTM
Biographical Dictionary of Biologists
http://www.cshl.org/comfort/scientists
The Faces of Science: African Americans in the Sciences
http://www.lib.lsu.edu/chem/display/faces.html
Biographical Sketches of Indian Scientists
http://www.acsu.buffalo.edu/~gupta/biography.html
Nobel Prize Internet Archive
http://www.nobelprizes.com
Women of Science at the Marine Biological Laboratory
http://www.mbl.edu/html/WOMEN/intro.html
Women in Developmental Biology
http://mit.edu/afs/athena.mit.edu/org/w/womens-studies/www/dev-bio/

Chapter 1

Multidisciplinary Sources

DICTIONARIES, ENCYCLOPEDIAS, AND COLLECTIONS OF PROFILES

American Men and Women of Science 1995-96. 19th ed. 8 vols. New Providence, N.J.: R. R. Bowker, 1994.
A primary research tool for information on active American scientists and engineers since the first edition appeared in 1906. The 123,406 persons in the nineteenth edition were chosen for distinguished achievement, research activity of high quality, or high rank or substantial responsibility in a scientific organization or research effort, particularly in the physical and biological sciences. Entries list each scientist's personal, educational, and career data as well as address, fax number, and e-mail address. Volume 8 contains charts of winners of various prizes, such as the Nobel Prize, and an index listing scientists by specialties and by state within each specialty.

Arnold, Lois Barber. *Four Lives in Science.* New York: Schocken Books, 1984.
Arnold furnishes biographical essays on Marie Martin Bachman (1796-1863), Almira Hart Lincoln Phelps (1793-1884), Louisa C. Allen Gregory (1848-1920), and Florence Bascom (1862-1945) in order to portray the difficulties women had when entering the male-dominated profession of science in nineteenth-century America. Accordingly, Arnold opens with an essay tracing the development of American scientific education and women's roles in it. In the biographical essays, she

focuses on the education and careers of her subjects, although she mentions their achievements in science. A final chapter summarizes her findings about American scientific education based upon her subjects' experiences. Photos; end notes; index. 179 pp.

Asimov, Isaac. *Asimov's Biographical Encyclopedia of Science and Technology.* 2d ed. New York: Doubleday, 1982.
A one-man effort, this collection has 1,510 scientists, ancient and modern, from the physical and life sciences. Asimov arranged the entries chronologically and numbered them. Each is a capsule biography, emphasizing the scientist's major achievements, and has cross-references to other scientists involved in the same work. The alphabetized table of contents refers to entries by number. Photos and drawings of scientists; index. 941 pp.

Bailey, Martha J. *American Women in Science: A Biographical Dictionary.* Santa Barbara, Calif.: ABC-CLIO, 1994.
Representing primarily the natural sciences, the women in this collection all began their careers before 1950. Bailey drew from the first three editions of *American Men and Women of Science* (see listing above). She also selected women who gained recognition in their fields (as, for instance, by election to the National Academy of Sciences) or because they made scientific contributions in government agencies or private organizations. Otherwise, she says, the criteria for selection were "problematic." The entries, more than 400, begin with headings listing education, employment record, and marriage. A 200-300-word essay then describes the personal background and professional career of each woman. A list of biographical sources ends each entry. Photos; bibliography; index. 463 pp.

Bernstein, Jeremy. *Experiencing Science.* New York: Basic Books, 1978.
Articles that Bernstein wrote for *The New Yorker* and other periodicals. The subjects are Johannes Kepler, I. I. Rabi, Trofim Denisovich Lysenko, Rosalind Franklin, and Lewis Thomas. (Science fiction author Arthur C. Clarke is also featured.) Superbly written, the articles combine personal biography

with science explication; the articles on Rabi and Thomas have long passages of quotations, recorded during extensive interviews. Brief biography; index. 275 pp.

The Biographical Dictionary of Scientists. 2d ed. New York: Oxford University Press, 1994.
This collection features astronomers, life scientists, chemists, engineers, geologists, mathematicians, and physicists from throughout the ancient and modern worlds. The editors selected the men and women responsible for the greatest scientific achievements, and the entries are intended to describe their personal lives and not just their status as scientists. The entries place scientists in their national and religious cultures as well as identifying their education, institutional affiliations, and discoveries. Major publications are mentioned in the short essays as well. A substantial essay, "Historical Review of Sciences," prefaces the collection. Some illustrations and graphs; glossary; index. 891 pp.

The Biographical Encyclopedia of Science. 4 vols. New York: Marshall Cavendish, 1997.
Intended for junior and senior high school students, this set features 550 scientists from the natural sciences. The scope is international and extends to classical Greece. The editors made every effort to include women, minorities, and young scientists who have not appeared in such a reference work before, although well known in their fields. A special feature of this set is its separation of a biographical essay for each scientist (500-700 words) from accompanying sidebars (250-500 words, from one to three) explaining the principal achievements of the scientist. Head matter containing vital statistics and bibliographies accompanies the alphabetized essays. Photos; indexes.

Biographical Memoirs. 69 vols. Washington, D.C.: National Academy Press, 1877-1996 .
This series of volumes contains obituaries, short biographies, and selected bibliographies of members of the National Academy of Sciences. It is an invaluable resource about prominent scientists in United States history. The essays, by

colleagues of the subjects, are brief, outlining the personal life and commenting on the career of each subject. Much of the information, especially the evaluations of careers, can be found in no other publication. Photos; cumulative index in volume 65.

Cohen, Harry, and Itzhak J. Carmin. *Jews in the World of Science.* New York: Monde Publishers, 1956.
This volume opens with a warm tribute to Albert Einstein by J. Robert Oppenheimer and essays on Einstein, Chaim Weizmann, Jonas Salk, and leading scientists of Jewish heritage in general. The main section lists in alphabetical order more than 3,000 Jewish scientists from throughout history. The brief entries give vital statistics, career information, organizational memberships, and prizes won. The coverage extends to scientists of all disciplines, engineers, and physicians, as well as some science educators. Photos. 263 pp.

Collins Biographical Dictionary of Scientists. 4th ed. Glasgow, Scotland: HarperCollins, 1994.
The coverage extends from classical antiquity to the present and is international. The approximately 900 brief entries list personal data and then describe the education, career, and scientific achievements of each scientist. The collection concentrates on the natural and physical sciences, although some engineers are included. Appendices list Nobel laureates and biographical references. Indexes. 602 pp.

Crowther, J. G. *Men of Science.* New York: W. W. Norton, 1936.
The author profiles five scientists crucial to the development of science and industry in Britain during the early and mid-nineteenth century: Humphry Davy (1778-1829), Michael Faraday (1791-1867), James Prescott Joule (1818-1889), William Thomson (Lord Kelvin, 1824-1907), and James Clerk Maxwell (1831-1879). Crowther covers the personal lives as well as the careers of the five men and clearly explains the ideas involved for readers with a basic understanding of chemistry and physics. Photos and drawings; short bibliographies for each chapter; index. 332 pp.

Crowther, J. G. *Scientists of the Industrial Revolution.* London: Cresset Press, 1962.
Crowther chronicles an era of British science through extended portraits of its four most influential figures: Joseph Black (1728-1799), James Watt (1736-1819), Joseph Priestley (1733-1804), and Henry Cavendish (1731-1810). According to Crowther, these four were responsible for the industrialization of science, which increased Great Britain's wealth and expanded its international influence enormously. The style is entertaining and intended for general readers, and Crowther discusses both the personal life and the career of each man. Portraits; end notes; index. 365 pp.

Crowther, J. G. *Statesmen of Science.* London: Cresset Press, 1965.
Crowther writes about the men who redirected British scientific fields in the nineteenth century—including astronomy, natural history, physics, chemistry, and biology—and in the process made science more practical for society and nurtured the late Industrial Revolution and early twentieth-century technology. Not all of the men profiled were scientists; one is Prince Albert, Queen Victoria's husband. The scientists are William Robert Grove (1811-1896), William Cavendish (1808-1891), Richard Burdon Haldane (1856-1928), Henry Thomas Tizard (1885-1959), and Frederick Alexander Lindemann (Lord Cherwell, 1886-1957). In each case, Crowther emphasizes the political and industrial work done on behalf of British science, although he mentions the research of the scientists among the nine. Photos; end notes; index. 391 pp.

Daintith, John, Sarah Mitchell, Elizabeth Tootill, and Derek Gjertsen. *Biographical Encyclopedia of Scientists.* 2 vols. Bristol, England: Institute of Physics Publishing, 1994.
This set contains about 2,000 "important scientists" from antiquity to the present, primarily in physics, chemistry, biology, astronomy, and the earth sciences; however, some physicians, mathematicians, engineers, anthropologists, sociologists, and philosophers are included. Inclusion, in all cases, derives from the person's having made a major discovery or advance in theory. Entries have two parts. The first lists biographical data. The second is an essay that focuses on the per-

son's scientific achievements. Appendices provide a chronology of scientific discoveries and major publications, brief histories of scientific institutions of note, and a list of influential publications with short descriptions of their contents. Drawings and graphs; indexes.

Dictionary of Scientific Biography. 18 vols. New York: Charles Scribner's Sons, 1970-1990.
The American Council of Learned Societies sponsored this authoritative collection of more than 7,000 articles, written for the most part by academic scholars. The articles cover scientists from classical antiquity to modern times, although no scientist living at the time of a volume's publication is included. Because most scientists worked in the biological sciences until the late nineteenth century, about one-third of the articles concern these fields; the editors also include mathematicians, physicists, astronomers, chemists, Earth scientists, and some historians and philosophers of science. The articles, which are sometimes very substantial, emphasize subjects' scientific achievements and careers. There is minimal personal biography. A bibliography of works about the scientist concludes each article. The editors note that women are underrepresented, which they intend to correct in future volumes. In general, readers with a high school education can understand the text. Indexes in vols. 16 and 18.

Elliott, Clark A. *Biographical Dictionary of American Science.* Westport, Conn.: Greenwood Press, 1979.
Elliott intends this collection to supplement *American Men and Women of Science* (listed above). He includes 600 major articles of about 300 to 400 words each and 300 minor articles of about 50 words each; all are very condensed, designed for reference only. Elliott includes figures in mathematics, astronomy, physics, chemistry, botany, geology, and some related disciplines during the seventeenth through nineteenth centuries. The major articles list a subject's birth and death, parentage, marital status, education, honors, career, society memberships, scientific contributions, published works, manuscripts, and biographical works. The minor articles abbreviate the information. Index. 360 pp.

Elliott, Clark A. *Biographical Index to American Scientists: The Seventeenth Century to 1920.* Westport, Conn.: Greenwood Press, 1990.

A supplement to Elliott's *Biographical Dictionary of American Scientists* (see entry immediately above). For approximately 2,850 American scientists who died before 1921, Elliott lists basic information about birth and death, field, occupation, manuscripts, and biographical material in other collections. The fields covered are mathematics, astronomy, physics, chemistry, geology, botany, zoology, and related specialties. A few anthropologists and psychologists also appear in the collection. Index. 300 pp.

The Excitement and Fascination of Science. 3 vols. Palo Alto, Calif.: Annual Reviews, 1965-1990.

These volumes collect philosophical and autobiographical essays by scientists, reprinted from various professional journals. The primary readership is other scientists, but the essays are treasures for biographers and historians. The first volume has 35 memoirs from annual reviews of biochemistry, pharmacology, chemistry, and physiology, and among the contributors are Linus Pauling, Eugene F. Dubois, W. F. Libby, and Otto Warburg. The second volume adds essays from the annual reviews of anthropology, astronomy and astrophysics, biochemistry, Earth and planetary scientists, entomology, fluid mechanics, genetics, microbiology, phytopathology, and plant physiology; its 34 essays include contributions by John C. Eccles, E. J. Öpik, and Georg von Békésy. The last volume, published in two parts, further adds reviews of immunology, neuroscience, nuclear and particle physics, nutrition, biophysics, and sociology; the two parts have 112 essays, and among the writers are Jesse L. Greenstein, Konrad Bloch, Fred L. Whipple, Joshua Lederberg, and Emilio Segrè. Cumulative index in vol. 3.

Gregory, Richard. *British Scientists.* London: Britain in Pictures, 1941.

Primarily a picture book, featuring color or black-and-white portraits, engravings, and drawings of thirty-one scientists. Roger Bacon is the earliest (1214-1292) and Oliver Lodge the most recent (1851-1940). The text provides superficial information about their scientific discoveries. The purpose is to celebrate British achievements. 47 pp.

Haber, Louis. *Black Pioneers of Science and Invention.* New York: Harcourt, Brace and World, 1970.
Haber writes to supply information about African American researchers in the United States, who were largely neglected in earlier biographical collections. Writing for a popular audience, Haber recounts the lives, work, and critical reception in white-dominated science of seven inventors and seven scientists, all men, from the eighteenth century to the mid-twentieth century. The scientists are agricultural chemist George Washington Carver, chemist Percy Lavon Julian, chemist Lloyd A. Hall, marine biologist Ernest Everett Just, surgeon Daniel Hale Williams, physician Louis Tompkins Wright, and physician Charles Richard Drew. The profiles equally discuss the subject's difficulties achieving recognition and the nature of his inventions or scientific discoveries. Photos and drawings; large bibliography; index. 181 pp.

Haber, Louis. *Women Pioneers of Science.* New York: Harcourt Brace Jovanovich, 1979.
Following an introductory essay about women in science, Haber offers short biographies—about 2,500 words each—on 12 women who made important contributions to areas of science, which include public health, education, psychology, biochemistry, physics, medicine, and biology: Alice Hamilton (1869-1970); Florence Rena Sabin (1871-1953); Lise Meitner (1878-1968); Leta S. Hollingworth (1886-1939); Rachel Fuller Brown (b. 1898); Gladys Anderson Emerson (b. 1903); Maria Goeppert Mayer (1906-1972); Myra Adele Logan (1909-1977); Dorothy Crowfoot Hodgkin (b. 1910); Jane C. Wright (b. 1920); Rosalyn S. Yalow (b. 1921); and Sylvia Earle Mead (b. 1935). The explanation of their scientific work is suited to general readers. Photos; bibliography; index. 171 pp.

Hackman, W. D. *Apples to Atoms: Portraits of Scientists from Newton to Rutherford.* London: National Portrait Gallery, 1986.
The book features reproductions of the portraits of 45 scien-

tists from the early seventeenth century to the late twentieth century (Stephen W. Hawking, for example). A few are color reproductions of paintings, but most are black and white. Accompanying each portrait is a sketch including the basic biographical information, scientific contribution, and major publications of the subject, as well as comments about the portrait, all directed at a general readership. Also illustrations of scientific phenomena and equipment; bibliography. 88 pp.

Hammond, Allen L., ed. *A Passion to Know*. New York: Charles Scribner's Sons, 1984.
Twenty profiles of nineteenth- and twentieth-century natural and social scientists written by popular science writers for a general audience. They are unusually lively, stylish profiles that fulfill the editor's purposes for the book: to show that the typical mass-media image of scientists is jejune and that the geniuses of modern science share a passionate curiosity for their work but otherwise show a full range of human personalities, philosophies, and quirks. Many standard figures are included, such as Charles Darwin and Albert Einstein, but most are seldom featured in popular press books, such as geneticist Barbara McClintock, medical researcher Carleton Gajdusek, physicist Frank Oppenheimer (J. Robert's brother), and biologist Brigit Zipser. An engaging introduction to modern science and scientists for readers with no science background. Photos; bibliography; index. 240 pp.

Harré, R., ed. *Some Nineteenth Century British Scientists*. Oxford, England: Pergamon Press, 1969.
This volume discusses eight scientists who represent the science of the late Victorian era: Charles Wyville Thomson and James Murray (both naturalists on the *Challenger* expedition, which circumnavigated the globe, beginning in 1872); mathematician Arthur Cayley; eugenicist Francis Galton (Charles Darwin's cousin); physicist William Thomson (Lord Kelvin); astrophysicist Norman Lockyer; metallurgist Sidney Gilchrist Thomas; and chemist William Ramsay. Most of the profiles are technically sophisticated, some containing advanced mathematics, and written in a formal style. Photos; short bibliographies for each chapter. 259 pp.

Hilts, Philip J. *Scientific Temperaments: Three Lives in Contemporary Science*. New York: Simon and Schuster, 1982.
The three portraits in this book attempt to connect the personality of each scientist to his style and success in science. Robert Wilson, the builder and administrator of the Fermi National Accelerator Laboratory, grew up as a cowboy and carried over his relentless self-reliance into physics and administration. Harvard geneticist Mark Ptashne, a bold and eccentric researcher, helped develop recombinant DNA technology. Shy and awkward in social settings, Joseph McCarthy thrived in working with computers, computer software, and artificial intelligence. The lives of these three, to Hilts, illustrate how a scientific temperament grows, finds its place, and creates. Based on extensive interviews, the profiles are gracefully written—Hilts has a flair for descriptive analogies—and insightful. End notes; index. 302 pp.

Holmyard, E. J. *British Scientists*. New York: Philosophical Library, 1951.
The author summarizes the scientific discoveries of 21 British scientists, beginning with Roger Bacon (1214-1294) and ending with William Ramsay (1852-1916), although he also has a chapter on investigations into atomic structure, which contains discussions of Joseph John Thomson (1856-1940) and Ernest Rutherford (1871-1937). The brief chapters devoted to the scientists contain little personal information other than career histories. A concluding chapter identifies British scientific societies. Drawings; bibliography. 88 pp.

Hutchings, Donald. *Late Seventeenth Century Scientists*. Oxford, England: Pergamon Press, 1969.
Written in the style of professional historians, the six chapters of this book each discuss the life and, especially, the work of a seventeenth-century savant: Robert Boyle, an English chemist; Marcello Malpighi, an Italian anatomist; Christopher Wren, an English architect; Christiaan Huygens, a Dutch physicist and mathematician; Robert Hooke, an English scientist of broad interest in physics and astronomy; and Isaac Newton. According to the editor, they were among the first generations of natural philosophers to favor experiment and rigorous logic

instead of metaphysical speculation. Photos and drawings; short bibliography for each chapter. 183 pp.

Hypatia's Sisters: Biographies of Women Scientists Past and Present. Seattle, Wash.: Feminists Northwest, 1976.
Brief sketches of 17 women natural and behavioral scientists written for a high-school level reading ability. Chosen from all over the world and from 300 B.C.E. to modern times, the women include Emilie du Châtelet, Mary Somerville, Florence Nightingale, Marie Curie, and Rachel Carson. An appendix lists twenty-three more women scientists and provides summary information. Drawings; end notes for each biographical sketch. 72 pp.

Jaffe, Bernard. *Men of Science in America.* New York: Simon and Schuster, 1958.
Jaffe supplies biographical sketches of 20 American scientists and the historical background necessary to understand their work. The first is the naturalist Thomas Harriot, who came to America with Walter Raleigh's 1588 expedition; the last is the Italian-born physicist Enrico Fermi. Between them are many famous names—Benjamin Franklin, Louis Agassiz, and Josiah Willard Gibbs, for example—as well as some not well known outside their specialty—such as the chemist Thomas Cooper and the entomologist Thomas Say. Jaffe wrote for general readers in order to help Americans appreciate their own scientific heritage. The sketches are indeed enjoyable reading, full of biographical detail and free of technical explanations that require training in science to understand. Photos, drawings, and diagrams; bibliography; index. 715 pp.

Jones, Bessie Zaban, ed. *The Golden Age of Science.* New York: Simon and Schuster, 1966.
This book collects 30 portraits of nineteenth-century scientists written by contemporaries. The coverage is international and includes mathematicians, physicists, astronomers, biologists, and chemists, beginning with England's William Herschel (1738-1822) and ending with Sweden's Svante Arrhenius (1859-1927). The essays vary in approach and style, but they all show how the scientists were preceived in their own eras.

The treatment is frequently technical but not beyond the grasp of readers with a basic knowledge of the sciences. 659 pp.

Kass-Simon, G., and Patricia Farnes, eds. *Women in Science: Righting the Record*. Bloomingdale: Indiana University Press, 1990.
Biography-history. In this book's introduction, Kass-Simon argues that the traditional disposition of society to honor the accomplishments of men, even if later proven wrong-headed, and to ignore those of women has long left women to function outside of history, unrecorded and unappreciated. This volume tries to right the record by showing to scholars and lay readers alike that far more women have contributed to all the sciences than typical biographical collections indicate. The ten main essays—by university professors for the most part—concern women in archaeology, geology, astronomy, mathematics, engineering, physics, biology, medical research, chemistry, and crystallography. Readers unaccustomed to feminist history may be put off by the seemingly offended tone that the authors take when discussing the suppression of women's efforts by men, but this book collects so much fresh information about the history of science that all readers should find it as enlightening as it is provocative. Photos; end notes for each essay; index. 398 pp.

Kessler, James H. *Distinguished African American Scientists of the 20th Century*. Phoenix, Ariz.: Oryx, 1995.
Written for middle and high school readers, this collection features 100 scientists and engineers, some not found in most other publications. Vital statistics preface essays that are three to five pages in length and recount subjects' upbringing, education, and career. Photos, drawings, and computer graphics; bibliographical references; index. 384 pp.

Lives in Science. New York: Simon and Schuster, 1948.
This "somewhat random sampling" from their magazine, according to the editors of *Scientific American*, shows men who dedicated themselves to science, among whom are "two or three authentic giants, a hero or two, a saint, and a rascal." There are 18 portraits in all, arranged in two chapters on cos-

mology and one each on fire, life, electromagnetism, and mathematics. The profiles are suitable for general readers and entertaining without sacrificing depth in explaining scientific ideas. Most of the subjects are predictable, such as Isaac Newton and Charles Darwin, but the mathematicians, at least, are surprising and intriguing choices: Charles Babbage, Lewis Carroll (that is, C. L. Dodgson), and Srinivasa Ramanujan. Drawings; bibliography. 274 pp.

McGraw-Hill Modern Scientists and Engineers. 3 vols. New York: McGraw-Hill, 1980.

The scope of this collection is international and concerns leading scientists from the 1920's through the Nobel Prize winners of 1978. Some of the articles are autobiographical, and many others were proofread by the subjects themselves. The intent of the volume is to give the reader, as far as possible, firsthand accounts in order to reveal how successful scientists found solutions to perplexing problems. Drawings; index.

McGrayne, Sharon Bertsch. *Nobel Prize Women in Science: Their Lives, Struggles, and Momentous Discoveries.* New York: Birch Lane Press, 1993.

McGrayne considers the answer to a single question throughout this fine collection of profiles: Why have so few women won Nobel Prizes in science? Her short biographies therefore tend to emphasize the struggles of her subjects, but she also presents the basic science involved for general readers. Not all of the subjects actually won a Nobel Prize. Marie Curie, Gerty Radnitz Cori, Irène Joliot-Curie, Barbara McClintock, Maria Goeppert Mayer, Rita Levi-Montalcini, Dorothy Crowfoot Hodgkin, Gertrude B. Elion, and Rosalyn Sussman Yalow did get prizes. Lise Meitner, Emmy Noether, Chien-Shiung Wu, Rosalind Franklin, and Jocelyn Bell Burnell did not, although, according to McGrayne, they did Nobel Prize-caliber work for which others won credit. Photos and diagrams; end notes; index. 419 pp.

McKissack, Patricia, and Frederick McKissack. *African-American Scientists.* Brookfield, Conn.: Millbrook Press, 1994.

For readers 12 years and older. For a long time, the

McKissacks write, bright young African Americans were discouraged from entering the sciences; because there were so few, it was thought African Americans were unfit for the sciences. Nevertheless, some ignored this rationalizing circular reasoning and had celebrated careers: the McKissacks profile these pioneering men and women. Among them are astronomer and almanac maker Benjamin Banneker (1731-1806), physician James Derham (b. 1762), marine biologist Ernest Everett Just (1883-1941), agricultural researcher George Washington Carver (1864-1943), chemist Percy Julian (1899-1975), and physicist Shirley Ann Jackson (b. 1946). Well written and inspirational for a young audience. Photos and drawings; bibliography; index. 96 pp.

McMurray, Emily J., ed. *Notable Twentieth-Century Scientists*. 4 vols. New York: Gale Research, 1995.
This set covers about 1,300 scientists in astronomy, biology, botany, chemistry, Earth science, environmental science, ecology, computer science, mathematics, medicine, physics, technology, and zoology. Among them are about 225 women and 150 minority scientists. For each entry vital statistics precede a sketch, ranging in length from 400 to 2,500 words, and a bibliography. The essays focus on the achievements and awards that gave scientists prominence, although the entire career of each is outlined. Photos; indexes.

Meadows, Jack. *The Great Scientists*. New York: Oxford University Press, 1987.
Meadows uses short biographies of scientists to present the core of science history in the West before the quantum physics revolution. The scientists are Aristotle, Galileo Galilei, William Harvey, Isaac Newton, Antoine Lavoisier, Alexander von Humboldt, Michael Faraday, Charles Darwin, Louis Pasteur, Marie Curie, Sigmund Freud, and Albert Einstein. Richly illustrated, the text interlards the biographies with captions and sidebars explaining politics, institutions, technology, and controversies. It is an entertaining book to browse for readers high school age and older. Photos, drawings, and graphics; bibliography; glossary; index. 256 pp.

Millar, David, Ian Millar, John Millar, and Margaret Millar. *Chambers Concise Dictionary of Scientists.* Edinburgh, Scotland: W. and R. Chambers and the Press Syndicate of the University of Cambridge, 1996.

The brief essays in this volume cover the vital statistics, career, and major contributions of selected scientists. They are those "whose names are famous" in the natural sciences, space science, and mathematics, as well as some who were explorers, engineers, physicians, or surgeons. The coverage is international and from throughout history. A list of Nobel Prize winners and a chronology of major events in science follow the essays. Portraits; index. 461 pp.

Muri, Hazel. *Larousse Dictionary of Scientists.* New York: Larousse, 1994.

The coverage, 2,200 entries, includes researchers in the natural and physical sciences from throughout history, along with some earlier philosophers important to the development of the scientific method. Entries feature a short biographical sketch, vital statistics, and a list of accomplishments. The book provides thorough cross-referencing and appendices listing Nobel Prize winners. Index. 595 pp.

Ogilvie, Marilyn Bailey. *Women in Science.* Cambridge, Mass.: MIT Press, 1986.

This volume has two major sections: a biographical dictionary and an annotated bibliography of information sources. In the first section, Ogilvie includes 185 women of Western countries from antiquity to about 1910. Entries have two parts. First comes a listing of the vital statistics, nationality, specialty, parentage, professional positions, spouse, children, and sources of biographical information for each woman. Second is a short essay about her career and contribution to science. The criteria for selection admit women who lived before the twentieth century, when science careers were almost exclusively for men. Often these women worked in quasi-scientific professions, such as midwifery. The second section includes books and articles, but the annotations are minimal. A substantial essay, "Science and Women: A Historical Overview," prefaces the biographical section. Index. 254 pp.

Olby, R. C., ed. *Late Eighteenth Century European Scientists.* Oxford, England: Pergamon Press, 1966.
This collection features seven scientists who, according to the editor, exemplify the astonishing progress in electricity, astronomy, botany, and, especially, chemistry during the late eighteenth century. Selecting experimenters rather than theorists, the authors trace the relation between science, technology, and society as well as discuss the lives and works of their subjects: biologist Jean Lamarck, botanist Joseph Koelreuter, chemist Antoine Lavoisier, chemist Henry Cavendish, physicist Alessandro Volta, physicist James Watt, and astronomer William Herschel. Valuable for those interested in the history of science, this little book is written in the formal style of academic historians. Drawing and diagrams; brief bibliographies for each chapter. 209 pp.

Parry, Albert. *The Russian Scientist.* New York: Macmillan, 1973.
In the introduction, Parry points out that Americans know very little about Russia's scientists, even those who made fundamental discoveries. He tries to remedy that failing with his book, which he aims at general readers. He profiles physicist Mikhail Lomonosov, mathematician Nikolai Lobachevsky, chemist Dmitry Mendeleyev, pathologist Ilya Metchnikov, physiologist Ivan Pavlov, space visionary Konstantin Tsiolkovsky, and physicist Peter Kapitsa, thereby providing glimpses of Russian science from the eighteenth century through the twentieth century. A concluding section discusses science in the Soviet Union during the 1950's and 1960's. Photos; bibliography; index. 196 pp.

Prominent Scientists. 2d ed. New York: Neal-Schuman Publishers, 1985.
The 2,211 contemporary scientists in this collection are drawn from the natural sciences and engineering. The information is spare: entries cite vital statistics and specialties only; for further information they provide abbreviated references to collected biographies, science histories, works that include biographical sketches, and publications between 1960 and 1983. The reference abbreviations are explained in a separate bibliographical section, "Key to Books Indexed." 356 pp.

Sammons, Vivian Ovelton. *Blacks in Science and Medicine.* New York: Hemisphere Publishing, 1989.
The author, a librarian, presents the results of painstaking labors to fill a gap in the biography of African Americans. She has collected information about approximately 4,000 scientists in all fields, medical professionals, and engineers from throughout American history, although those who earned degrees following World War II predominate. Each entry lists the vital statistics, specialty, education, employment history, and organization memberships and provides references to information sources. Sammons also indicates if the subject was the first African American in a given specialty or in a region of the United States, as well as the subject's research interests. The listing is alphabetized, and names are grouped by specialty at the end of the volume. Bibliography. 293 pp.

Sarton, George. *Six Wings.* Bloomington: University of Indiana Press, 1957.
More history than biography, this book nevertheless contains much biographical information about Renaissance scientists—that is, those living from 1450 to 1600 in the West. The title reflects a popular Renaissance trope, derived from the Bible, for divisions of knowledge. Accordingly, Sarton divides his discussion into explorers and educators; mathematicians and astronomers, such as François Viète, Nicolaus Copernicus, and Tycho Brahe; physicists, chemists, and inventors, such as William Gilbert and Paracelsus; researchers in natural history, such as Georgius Agricola and Bernard Palissy; anatomists and physicians, such as Andreas Vesalius; and "Leonardo da Vinci: Art and Science." The book is based on a lecture series for college students. Portraits; end notes; index. 318 pp.

Schwartz, George, and Philip W. Bishop. *Moments of Discovery.* 2 vols. New York: Basic Books, 1958.
This anthology is only secondarily biographical. The editors have brought together excerpts from the writing of the great scientists of Western history, and many famous names in the natural sciences are included (with the exception of Marie Curie it is an exclusively male group). The editors preface each excerpt with a short biographical essay, primarily about

the scientist's career. The first volume covers classical times through the Renaissance, including Isaac Newton. The second volume is for modern science, which the editors believe is characterized by increasing specialization. Accordingly, they divide the entrants in sections devoted to anatomy, biology, evolution, medicine, chemistry, and physics. The only section that is underrepresented is the last, physics, which concludes with J. Robert Oppenheimer but has nothing from Albert Einstein, Niels Bohr, or Enrico Fermi. Index.

Simmons, John. *The Scientific 100: A Ranking of the Most Influential Scientists, Past and Present*. Secaucus, N.J.: Citadel Press, 1996.
Simmons includes scientists and philosophers from all social and natural sciences and all periods in the West and introduces them in order of their influence on later scientists and society. His choices are bound to surprise and antagonize every reader somewhere along the list, but that is part of the fun of such a book—to give readers a chance to disagree. Few will object to finding Isaac Newton listed first, or Albert Einstein second, but many may wonder why Aristotle does not earn a rank, or why Trofim Lysenko (number 93), the destroyer of Soviet genetics, appears at all. Franz Boas (14) before Charles Lyell (29)? In each case Simmons defends his rankings with historical arguments, although he admits some of the choices are a matter of taste and viewpoint. A fascinating book that combines biographical information and science history. Photos and drawings; end notes; bibliography; index. 504 pp.

Stille, Darlene R. *Extraordinary Women Scientists*. Chicago, Ill.: Children's Press, 1995.
An exceptionally comprehensive collection devoted to readers 12 years and older, this volume sketches the lives and describes the scientific achievements of 50 women and has prefactory essays about modern women, scientists in ancient times, mathematicians, physicians, and inventors and engineers. Arranged alphabetically, the subjects are a hall of fame of women scientists and include researchers in all disciplines, such as astronomer Annie Jump Cannon, ecologist Rachel Carson, zoologists Jane Goodall and Dian Fossey,

biologist Barbara McClintock, chemist Marie Curie, physicists Lise Meitner and Maria Goeppert Mayer, geologist Florence Bascom, and anthropologist Margaret Mead. The passages about science are brief and require only the most rudimentary acquaintance with science to understand, and the profiles in general are written deftly, emphasizing the first times women entered a field or explored a topic. Photos; bibliography; index. 206 pp.

Turkevich, John. *Soviet Men of Science*. Princeton, N.J.: D. Van Nostrand, 1963.
The author tried to take some of the chill from the Cold War by introducing leading Soviet scientists to Americans. The scientists chosen, approximately 450, were all members of the Soviet Academy of Sciences, and Turkevich drew his information from Soviet literature. Each entry sketches the subject's personal and career biography, lists publications, and gives home and professional addresses. The coverage includes the natural sciences, mathematics, and engineering. 441 pp.

Turkevich, John, and Ludmilla B. Turkevich. *Prominent Scientists of Continental Europe*. New York: American Elsevier Publishing, 1968.
The authors list basic biographical, career, and publication data for more than 3,200 scientists in Europe, excluding the British Isles. They group the entrants by country; the coverage includes natural and social scientists, as well as physicians. Only members of national academies and professors at leading universities are included, and the listings draw from both questionnaires and secondary biographical sources. 204 pp.

Who's Who in Science and Engineering. 2d ed. New Providence, N.J.: Marquis-Who's Who, 1994.
The 23,600 entries include contemporary scientists, physicians, and engineers, mostly from the United States but with some representation of 115 other countries. The volume casts a wide net: It boasts coverage of 110 science specialties, including social sciences. The criteria restrict inclusion to those scientists who have won a prestigious award, directed important research, provided special leadership, been out-

standing members of honorary organizations (such as the National Academy of Sciences), or made remarkable contributions to their fields. The alphabetized entries list the occupation, vital statistics, parentage, marriage, children, education, professional certification, career information, publications, civic and political affiliation, military experience, awards and fellowships, association memberships, and address of each. An appendix lists the winners of various awards, including the Nobel laureates. Indexes. 1,269 pp.

Who's Who in Science in Europe. 3d ed. 4 vols. Guernsey, England: Francis Hodgson, 1978.
Nearly 50,000 entries and cross-references to Europeans in the natural, physical, medical, and agricultural sciences, as well as some in archaeology and architecture. Each entry lists the person's name, position, degrees, specialty, education, career information, association memberships, research interests, and address.

Wilkinson, Philip, and Michael Pollard. *Scientists Who Changed the World*. New York: Chelsea House, 1994.
For readers 12 years and older, this book offers three types of useful information. First, it lists prominent scientists from throughout the world and history, from Galileo Galilei in Renaissance Italy to Francis Crick, James Watson, Rosalind Franklin, and Maurice Wilkins and their decoding of deoxyribonucleic acid (DNA) in England in 1953; there are also chapters devoted to such famous figures as Johannes Gutenberg, the first printer, and the first men on the Moon. Second, the authors describe the basic nature of discoveries of the scientists but also include some, such as the invention of gunpowder, for which no single figure is credited. Third, its chapters ordered chronologically, the book provides an outline of the history of science. Drawings; bibliography; index. 93 pp.

Williams, Trevor I., ed. *A Biographical Dictionary of Scientists*. 2d ed. London: Adam and Charles Black, 1974.
This volume offers short, plainly written entries on approximately 1,000 scientists in the natural sciences, medicine, technology, and mathematics. Written by university professors,

including several eminent biographers, such as Angus Armitage and L. Pearce Williams, the book is intended for students and general readers. Following a short paragraph of vital statistics, each entry concentrates on the scientist's best-known discoveries or inventions. Appendices list births of scientists in chronological order and vital statistics of scientists mentioned in the book but not allotted entries of their own. A short bibliography for each entry. 641 pp.

Wilson, Grove. *The Human Side of Science.* New York: Cosmopolitan Book Corporation, 1929.
Biography-history. To understand "searchers after strange truths," Wilson writes, is to understand the best of human nature. His twenty-eight searchers are the most famous names of Western science from classical antiquity (Thales) to Albert Einstein. The list includes only chemists, physicists, physicians, astronomers, and biologists, with the exception of aeronautical engineer Samuel Pierpont Langley. Each biography mixes personal information with accounts of the scientist's career and achievements in order to emphasize the "human side" of the book's title. Pleasant, but the awestruck tone makes the book little more than inspirational reading. Photos; bibliography. 397 pp.

World Who's Who in Science. Chicago, Ill.: Marquis-Who's Who, 1968.
An alphabetical registry of nearly 30,000 scientists worldwide, from antiquity to the third quarter of the twentieth century. The emphasis is on the physical and biological sciences, but some social scientists are included. Based on staff research into historical records (and questionnaires sent back from scientists still living), the entries list the nationality, specialty, vital statistics, parentage, degrees, education, spouse, children, employment, honors, associations, research interests, and address of each. Bibliography. 1,855 pp.

Yost, Edna. *American Women of Science.* Philadelphia, Pa.: J. B. Lippincott, 1955.
This early collection of biographies of women scientists was to offer young readers realistic role models, but adults will find

it a pleasure to read as well. The 12 women chosen represent medicine, astronomy, zoology, biology, physics, and anthropology. Among them are Annie Jump Cannon (1863-1941), Florence Rena Sabin (1871-1953), Katherine Blodgett (b. 1898), and Margaret Mead (1901-1978). Although other collections supersede this one for comprehensiveness and depth, Yost writes with unusual freshness because she had not uncovered any earlier biographical collections dedicated to women and was surprised at how much in quality and variety American women had contributed to science. 233 pp.

SERIES FOR YOUNG READERS

Immortals of Science. New York: Franklin Watts, 1961-1968.

For readers 12 years or older, these full-length biographies not only recount scientists' lives but also describe the cultures that produced them. The authors explain scientific discoveries and theories at length. The series' titles are *Aristotle: Dean of Early Science* by Glanville Downey (1962); *Robert Boyle: Founder of Modern Chemistry* by Harry Sootin (1962); *The Curies and Radium* by Elizabeth Rubin (1961); *Michael Faraday and the Electric Dynamo* by Charles Paul May (1961); *William Harvey: Trailblazer of Scientific Medicine* by Rebecca B. Marcus (1962); *Johannes Kepler and Planetary Motion* by David C. Knight (1962); *Robert Koch: Father of Bacteriology* by Knight (1963); *Leonardo da Vinci: Pathfinder of Science* by Henry S. Gillette (1962); *James Clerk Maxwell and Electromagnetism* by May (1962); *Isaac Newton: Mastermind of Modern Science* by Knight (1961); *Hippocrates, Father of Medicine* by Herbert S. Goldberg (1963); *The Wright Brothers: Pioneers of Power Flight* by Carroll V. Glines (1968); *Archimedes and the Door of Science* by Jeanne Bendick (1962); *Alessandro Volta and the Electric Battery* by Bern Dibner (1964); *Joseph Priestley: Pioneer Chemist* by Marcus (1961); *Sigmund Freud* by Francine Klagsbrun (1967); *Carl Linnaeus: Pioneer of Modern Biology* by Alice Dickinson (1967); *Copernicus: Titan of Modern Astronomy* by Knight (1967); *Wilhelm Conrad Roentgen and the Discovery of X Rays* by Dibner (1968); *Philippe Pinel, Unchainer of the Insane*, by Bernard Macklin (1968); and *Albert Einstein and the Theory of Relativity* by Herbert Kondo (1968). All volumes come with indexes; some have glossaries of technical terms. About 110-150 pp.

Parker, Steve. Science Discoveries. New York: HarperCollins, 1992-1996.
For readers eight years and older. Simply written text, a variety of sidebars to accompany the main narrative, and an abundance of lovely illustrations make Parker's books eye-catching and entertaining for young readers—fine introductions to the world of scientists. They include *Charles Darwin and Evolution* (1992), *Galileo and the Universe* (1992), *Thomas Edison and Electricity* (1992), *Marie Curie and Radium* (1992), and *The Wright Brothers and Aviation* (1995). Many color illustrations and photos; glossary; index. About 30 pp.

Pioneers in Change. Englewood Cliffs, N.J.: Silver Burdett Press, 1989-1992.
For readers 12 years and older. The plainly written texts follow the careers of scientists and place them in their historical context. Scientific discoveries are explained in fairly general terms. Throughout, the emphasis is on showing how each scientist changed American life and thought. Titles include *Albert Einstein* (1989) by Karin Ireland; *Alexander Graham Bell* (1989) by Kathy Pelta; *Thomas Alva Edison* (1989) by Vincent Buranelli; *Margaret Mead* (1989) by Julie Castiglia; *Jane Addams* (1990) by Leslie A. Wheeler; *Jonas Salk* (1990) by Marjorie Curson; *Lewis Howard Latimer* (1991) by Glennette Tilley Turner; *John Muir* (1990) by Eden Force; *Buckminster Fuller* (1990) by Robert R. Potter; *George Washington Carver* (1991) by James Marion Gray; *The Wright Brothers* (1991) by Richard M. Haynes; *Benjamin Franklin* (1991) by Potter; *Robert H. Goddard* (1991) by Clafford Karin Farley; and *George Eastman* (1992) by Burnham Holmes. Each contains photos, bibliography, and index. About 110-140 pages long.

Pioneers of Science. New York: The Bookwright Press, 1990-1992.
For readers ten years and older, the well-illustrated books in this series focus on the lives of their subjects and the scientific discoveries, giving cursory attention to the historical setting. Titles include *Archimedes* by Peter Lafferty (1991); *Alexander*

Graham Bell by Andrew Dunn (1991); *Karl Benz* by Brian Williams (1991); *Marie Curie* by Dunn (1991); *Albert Einstein* by Lafferty (1992); *Michael Faraday* by Michale Brophy (1991); *Galileo* by Douglas McTavish (1990); *Guglielmo Marconi* by Nina Morgan (1991); *Isaac Newton* by McTavish (1990); *Louis Pasteur* by Morgan (1992); *Joseph Lister* by McTavish (1992); *Edward Jenner* by Stephen Morris (1992); and *Leonardo da Vinci* by Lafferty (1990). Photos and drawings; glossaries; short bibliographies; indexes. 48 pp.

Pioneers of Science and Discovery. London: Priory Press, 1973-1978.
For readers 12 years and older. Substantial treatments of well-known scientists and explanations of their most celebrated achievements or the scientific specialty they pioneered. Titles include *Edward Jenner and Vaccination* by Anthony J. Harding (1974); *Louis Pasteur and Microbiology* by Harold I. Winner (1974); *Alfred Nobel: Pioneer of High Explosives* by Trevor I. Williams (1974); *Michael Faraday and Electricity* by Brian Bowers (1974); *James Simpson and Chloroform* by Richard S. Atkinson (1973); *James Cook: Scientist and Explorer* by Williams (1974); *Ernest Rutherford and the Atom* by Philip B. Moon (1974); *Isaac Newton and Gravity* by Piyo Rattansi (1974); *Marie Stopes and Birth Control* by Harry V. Stopes-Roe (1974); *Alexander Fleming and Penicillin* by W. Howard Hughes (1974); *Gregor Mendel and Heredity* by Wilma George (1975); *William Harvey and the Circulation of the Blood* by Eric Neil (1975); *Albert Einstein and Relativity* by Derek J. Raine (1975); *Joseph Lister and Antisepsis* by A. J. Harding Rains (1977); *John Logie Baird and Television* by Michael Hallett (1978); and *Leonardo da Vinci and the Art of Science* by Kenneth D. Keele (1977). Photos and drawings; bibliography; index. About 96 pp.

Chapter 2

Astronomy and Cosmology

COLLECTIONS

Ball, Robert S. *Great Astronomers*. London: Isbister, 1901.
The publication date makes the author's choice of "great" astronomers sometimes unusual, since subsequent history has obscured the names of some eighteenth- and nineteenth-century men that are featured. A brief general introduction discusses ancient astronomy. Beginning with Ptolemy, Ball has biographical-scientific sketches of 18 astronomers; among them are Nicolaus Copernicus, Galileo Galilei, Johannes Kepler, Isaac Newton, Simon Laplace, and William and John Herschel. For these figures readers can easily find better accounts published more recently, but for astronomers like James Bradley, Jean Joseph Le Verrier, or John Couch Adams—all influential in their day—Ball offers rare information. Portraits. 372 pp.

The Biographical Dictionary of Scientists: Astronomers. New York: Peter Bedrick Books, 1984.
Following a historical sketch of astronomy, this volume has short articles on 205 natural philosophers, astronomers, astrophysicists, and cosmologists, as early as Thales of Miletus (ca. 624-547 B.C.E.). Entries average about 500 words each. They provide basic biographical information and a summary of each scientist's achievements for general readers. A valuable basic resource because of the extent of coverage and the brevity of the articles. Photos and drawings; glossary; index. 204 pp.

Christianson, Gale E. *The Wild Abyss*. New York: Free Press, 1978.
Biography-history. Christianson chronicles the development of modern astronomy by concentrating the narrative on the figures who transformed it into a scientific discipline. First he devotes three chapters to ancient and medieval conceptions of the universe. Then he profiles Nicolaus Copernicus, Tycho Brahe, Johannes Kepler, Galileo Galilei, and Isaac Newton. A well-written work of intellectual history for a general audience. Photos, drawings, and diagrams; end notes; bibliography; index. 461 pp.

Heath, Thomas. *Aristarchus of Samos: The Ancient Copernicus*. Oxford, England: Clarendon Press, 1913.
This scholarly tome contains two parts. The first part comprises portraits of early Greek astronomers, including Aristotle, Eudoxus, and Plato, and reviews their ideas. The second part has the Greek text and a translation of an astronomical treatise by Aristarchus of Samos (ca. 310-230 B.C.E.) concerning a heliocentric model of the universe. It is prefaced by a substantial sketch of his life and teachings. Readable but suffused with the ponderous tone of serious classical scholarship; readers should also know trigonometry to appreciate the explanations of astronomy. Footnotes; index. 425 pp.

Lightman, Alan, and Roberta Brawer. *Origins: The Lives and Worlds of Modern Cosmologists*. Cambridge, Mass.: Harvard University Press, 1990.
The authors provide a short biographical sketch for each of twenty-seven physicists and astronomers who have made major contributions to cosmology in the second half of the twentieth century, such as Fred Hoyle, Stephen W. Hawking, and Steven Weinberg. Then they present the results of interviews with each scientist in question-and-answer format, the questions pertaining to problems in modern theory as well as to biographical topics. A long introduction summarizes cosmological theory to prepare readers for the interviews. Photos; bibliography; glossary; end notes. 563 pp.

Lodge, Oliver. *Pioneers of Science*. New York: Macmillan, 1926; Dover, 1960.

Originally written in 1893, this is a classic in scientific biography used to illustrate an idea. Lodge's purpose is to trace the development of the scientific method through the lives and achievements of prominent astronomers, such as Nicolaus Copernicus, Galileo Galilei Isaac Newton, and William Herschel. There is more science exposition than biography. Pleasant reading for a general audience, although the science is often glaringly out of date and Lodge takes a strangely dim view of his contemporary scientists. Drawings; index. 404 pp.

MacPike, Eugene Fairfield. *Hevelius, Flamsteed and Halley: Three Contemporary Astronomers and Their Mutual Relations*. London: Taylor and Francis, 1937.
MacPike allots a chapter each to the lives and scientific discoveries of Johannes Hevelius, John Flamsteed, and Edmond Halley and then discusses how the three, in helping one another, changed astronomy. The author writes in hopes of clarifying the historical record of a crucial era for astronomy, and the book is closely argued, if labored. It contains much biographical information on the three men. Drawings; bibliography; index. 140 pp.

Overbye, Dennis. *Lonely Hearts of the Cosmos: The Scientific Quest for the Secret of the Universe*. New York: HarperCollins, 1991.
Biography-history. Overbye follows the development of twentieth-century cosmology through portraits of leading contributors, explanations of their work, and anecdotes about their feuds, told with relish. Principal coverage is of Allan Sandage, Stephen W. Hawking, James Peebles, Yakov Boris Zel'dovich, and David N. Schramm. The book covers the period from Edwin P. Hubble's discovery of the expanding universe to the recognition, during the 1980's, of large-scale bubbles in the cosmic structure. The narrative, intended for general readers, is richly descriptive, both of scientists' personalities and of physics, and frequently dramatic. Photos; index. 438 pp.

Richardson, Robert S. *The Star Lovers*. New York: Macmillan, 1967.
Written primarily to entertain lay readers, the sixteen chapters in this book sketch the lives and work of famous astronomers,

cosmologists, and astrophysicists, such as Tycho Brahe, Isaac Newton, Percival Lowell, Albert Einstein, and Walter Baade. Richardson hopes to inspire interest in astronomy in general and to capture something of the scientists' personalities. The coverage extends from 1550 to 1960, and the explanations of science are moderately sophisticated. Drawings and photos; bibliography; index. 310 pp.

Rosen, Edward. *Three Imperial Mathematicians: Kepler Trapped Between Tycho Brahe and Ursus*. New York: Abaris Books, 1986.
In this narrowly focused scholarly treatise, Rosen examines the relations among Nicolas Reimers (also known as Ursus), Tycho Brahe, and Johannes Kepler in order to establish how the modified form of Nicolaus Copernicus' cosmology was developed. It is a story of jealousy and vicious intrigue as the men successively win the post of mathematician to the Holy Roman Emperor. Rosen employs extensive quotations from the works of the three in telling the story, which has a wealth of biographical data as well as explanations of their various astronomical theories. Drawings; bibliography; end notes; index. 384 pp.

Swift, David W. *SETI Pioneers: Scientists Talk About Their Search for Extraterrestrial Intelligence*. Tucson: University of Arizona Press, 1990.
Swift offers short biographical sketches and question-and-answer interviews with 16 scientists who pioneered the search for extraterrestrial intelligence (SETI). Among the subjects are Philip Morrison, Frank Drake, Josef Shklovskii, Carl Sagan, and Freeman Dyson. The interviews deal mostly with SETI, even though the scientists all have other accomplishments and interests in astronomy, aerospace, and physics. Photos; bibliography; glossary; index. 434 pp.

BIOGRAPHIES AND AUTOBIOGRAPHIES

Adamczewski, Jan. *Nicolaus Copernicus and His Epoch*. Philadelphia, Pa.: Copernicus Society of America, 1973.
Biography. A richly illustrated, laudatory account of Nicolaus

Copernicus' life with a light treatment of his medical and cosmological ideas. Packed with historical information, this book offers a non-technical introduction to the late Renaissance and can serve as a basis for more detailed treatments of Copernicus' science. Photos, drawings, and diagrams. 161 pp.

Anderson, Margaret J. *Isaac Newton: The Greatest Scientist of All Time*. Springfield, N.J.: Enslow Publishers, 1996.
Biography for readers ten years and older. This book's subtitle conveys the general tone clearly. Anderson informs young readers of the life of Isaac Newton (1642-1727) from his obscure birth to his death, after he had been knighted and become powerful in scientific circles. She writes clearly about Newton's theory of gravity and discovery of the spectrum. She includes reference to Newton's quarrels with other scientists but for the most part avoids incidents that show him unfavorably. The text concludes with directions for experiments based upon Newton's ideas that readers can perform. Photos, drawings, and diagrams; bibliography; glossary; end notes; index. 128 pp.

Armitage, Angus. *Copernicus, the Founder of Modern Astronomy*. New York: A. S. Barnes, 1962.
Scientific biography. Armitage concentrates more on the cosmology of Nicolaus Copernicus (1473-1543) than he does in his earlier biography, *Sun, Stand Thou Still* (see below). In fact only about one-fifth of the book contains strictly biographical information. The rest addresses the composition and nature of *De revolutionibus*, the Copernican heliocentric model, the theory of the Moon's motion and planetary orbits, and the reception of the Copernican theory. It is all related in considerable mathematical detail, which requires a knowledge of geometry and trigonometry. Drawings and diagrams; bibliography; index. 236 pp.

Armitage, Angus. *Edmond Halley*. London: Nelson, 1966.
Scientific biography. His book, Armitage writes, is not a definitive biography but "an historical evaluation of Halley's scientific researches" with "some account of his life story."

Accordingly, Armitage prepares the reader with an introductory chapter outlining the rise of science in Europe before Halley's life (1656-1742). Halley was a man of wide interests and accomplishments. A friend of Isaac Newton, Halley served as master of the mint and Astronomer Royal and published research in mathematics, astronomy, atmospheric science, physics, geophysics, and demography. He was also an accomplished navigator and Arabist. Armitage's detailed explanations of Halley's thinking frequently incorporate geometry, trigonometry, and calculus. The portion of the book devoted to personal biography is sketchy. Photos, diagrams, and drawings; index. 220 pp.

Armitage, Angus. *John Kepler*. New York: Roy Publisher, 1966. Biography. Armitage covers the life and ideas of Johannes Kepler (1571-1630) primarily to show how his mathematical treatment of planetary relationships prepared the way for Isaac Newton's theory of gravity. But Armitage also discusses Kepler's interest in astrology, his observational work for Tycho Brahe, and his discovery of a supernova. The technicalities of Kepler's ideas receive careful exposition for readers who remember their high school geometry. Drawings and diagrams; glossary; index. 194 pp.

Armitage, Angus. *Sun, Stand Thou Still: The Life and Work of Copernicus the Astronomer*. New York: Henry Schuman, 1947. Biography-history. Armitage opens this book with a lengthy explanation of astronomy before Nicolaus Copernicus. Against this background, he relates Copernicus' life, especially his education, teaching career, observational work, and attempt at calendar reform. The last third of the book takes up Copernicus' heliocentric model of the universe and reactions to it from contemporary scientists and the Catholic church. Although this book does not address all Copernicus' ideas in detail (such as his medical ideas), it remains one of the clearest introductions to his life and times. Drawings and diagrams; bibliography; index. 210 pp.

Armitage, Angus. *William Herschel*. London: Thomas Nelson and Sons, 1962.

Scientific biography. Armitage allots only a short chapter to the personal life of William Herschel (1738-1822). The rest of the book concerns the astronomical discoveries of this self-taught genius and the ideas developed in his 70 scientific papers. Armitage organizes the material by subject, rather than chronologically, discussing the work on telescopes and solar physics, stars and parallax, and nebula. He concludes with an estimation of Herschel's pervasive influence on nineteenth-century astronomy. Readers will profit best from the book if they understand geometry and basic astronomy. Photos and diagrams; short bibliography; index. 158 pp.

Baumgardt, Carola. *Johannes Kepler: Life and Letters.* New York: Philosophical Library, 1951.
Biography. Baumgardt's translations of Johannes Kepler's letters occupy much of this book. She links them together with a biographical narrative concerned primarily with his personality and career. She does not explain Kepler's laws of motion in detail, nor his method of calculation, although in the introduction, Albert Einstein describes how Kepler calculated planetary orbits. Bibliography; index. 209 pp.

Bedini, Silvio A. *The Life of Benjamin Banneker.* New York: Charles Scribner's Sons, 1972.
Biography. Benjamin Banneker (1731-1806), the self-taught son of a slave, was one of the first African American scientists in the United States. As Bedini shows, Banneker's achievements were modest but solid, mainly in the publication of almanacs with astronomical and meteorological information. The author tries to give Banneker his due—a brilliant, hardworking man laboring in very difficult circumstances—while clearing away the errors and misstatements about Banneker's achievements made by other writers. The thoroughly researched text looks deep into Banneker's cultural and family background as well as his life. Photos and drawings; end notes; extensive bibliography; index. 434 pp.

Buttmann, Günther. *The Shadow of the Telescope: A Biography of John Herschel.* New York: Charles Scribner's Sons, 1970.
Biography. Buttmann tells the story of a son, John Herschel,

who had to labor in the shadow of a great father, William Herschel, who revolutionized modern astronomy. According to Buttmann, although the two men really do not compare well, John was as great an overall scientist. He conducted research on light, optics, spectral analysis, mathematics, electricity, magnetism, geology, and photography before taking over his father's survey of the heavens and cataloguing of double stars. Herschel was elected to the Royal Society at age 21, his scientific accomplishments grew continually, and he turned into the most accomplished British astronomer in the nineteenth century, just as his father had been in the eighteenth century. A brief but information-laden book. Photos and drawings; end notes; bibliography; index. 219 pp.

Caspar, Max. *Kepler*. New York: Abelard-Schuman, 1959; Dover, 1993.
Biography. The preeminent Kepler scholar of his day, Caspar presents him as a man of innate nobility who triumphed over misfortunes to become a figure worthy of "veneration and love." The rosy glow of this tone, however, does not obscure Caspar's extensive scholarship. He examines the intellectual milieu of Johannes Kepler's era, as well as the political influences on the scientist. The book includes a thorough but non-technical review of Kepler's major works and theories. Bibliography; index. 401 pp.

Chandrasekhar, S. *Eddington, the Most Distinguished Astrophysicist of His Time*. Cambridge, England: University of Cambridge Press, 1983.
Scientific biography. Subrahmayan Chandrasekhar summarizes the career of his teacher and intellectual adversary, Arthur Stanley Eddington (1882-1944). The two great scientists disagreed about the fate of large stars when they burn out (Chandrasekhar was right) but their disagreement hardly enters the narrative as Chandrasekhar explains Eddington's contributions to the understanding of stellar energy, relativity, stellar dynamics, and theoretical astrophysics. Readers must know advanced mathematics to understand the text fully. Photos and diagrams; end notes; index. 64 pp.

Christianson, Gale E. *Edwin Hubble: Mariner of the Nebulae.* New York: Farrar, Straus and Giroux, 1995.

Biography. A cautiously written study of Edwin P. Hubble (1889-1953), who has been called the greatest astronomer of the twentieth century. Certainly, his findings on the recession of galaxies and intergalactic distances provided an observational foundation to modern cosmology and made Hubble a celebrity, but in Christianson's deft account, Hubble, while a brilliant and exacting researcher, is far from an admirable figure. He, in fact, sounds like a distinctly American type, the self-made legend who rises from humble origins to earn international acclaim by sheer effort and, with American hubris, disdains his background. According to the author, Hubble exaggerated (if not fabricated) some of his early achievements, affected a British accent and manners, cultivated an Olympian demeanor, and turned his back on his Midwestern roots—all so he could live the role of the patrician intellectual. A fascinating life. Christianson explains the nature of Hubble's discoveries and the context from which they grew. Readers need no need prior knowledge of astronomy to appreciate the technical passages. Photos; bibliography; end notes; index. 420 pp.

Christianson, Gale E. *In the Presence of the Creator.* New York: Free Press, 1985.

Biography. Christianson's thesis is that everything the mature Isaac Newton did in his life accorded with his belief that he was one of God's elect and capable of penetrating the innermost secrets of nature. This captivating book explains Newton's achievements by locating them in the scientific and political culture of the seventeenth and eighteenth centuries. The reader learns about aspects of Newton's research that are often skipped over, such as his long investigations into alchemy and biblical exegesis. Christianson analyzes Newton's feuds with other scientists, his famous refusal to publish his discoveries, and his service as director of the mint. The style is rich in imagery, even ornate at times, as Christianson delivers an intellectually stimulating portrait for readers versed in the history of ideas and basic physics and mathematics. Photos; bibliography; end notes; index. 623 pp.

Christianson, Gale E. *Isaac Newton and the Scientific Revolution.* New York: Oxford University Press, 1996.
Biography for readers 14 years and older. Writing frankly and lucidly for his audience, Christianson does not santize Isaac Newton. He presents the irascible genius' quirks, feuds with other scientists, and obsessions. He also conveys Newton's world-view-changing theory of gravitation and discoveries in mathematics and optics and as clearly discusses his interest in alchemy and biblical interpretation. Christianson not only relates Newton's boyhood, professorship at Cambridge University, terms in Parliament, directorship of the mint, presidency of the Royal Society, and knighthood; he also explains the intellectual milieu of the times. Portraits and drawings; bibliography; index. 155 pp.

Drake, Stillman. *Galileo at Work.* Chicago, Ill.: University of Chicago Press, 1978.
Scientific biography. Drake discusses the scientific work of Galileo Galilei (1564-1642) in chronological order to understand the development of his ideas by tracing recurrent themes. His intention is to give a "fair overall view of Galileo's scientific career." Thus, he includes discussions of interests usually passed over by other biographers—such as the interest in music he inherited from his father—and translations of minor publications and letters. He does not provide a thorough portrait of Galileo's private life or comment upon his influence on later scientists. The text can be understood by attentive lay readers who understand geometry and basic physics, for the explanations of physical principles and astronomical discoveries are detailed. A helpful supplement to non-technical biographies of Galileo. Diagrams; appendix containing biographical information about contemporary scientists and mathematicians; end notes; index. 536 pp.

Fisher, Leonard Everett. *Galileo.* New York: Macmillan, 1992.
Biography for readers eight years and older. Writing clearly and simply, Fisher outlines the life and achievements of Galileo Galilei, dwelling on his defense of the Copernican model of the universe. The book's strong pointm is the text, although drawings dominate every page. Drawings. 27 pp.

Gade, John Allyne. *The Life and Times of Tycho Brahe.* Princeton, N.J.: Princeton University Press, 1947; New York: Greenwood, 1969.

Biography. Gade admits to knowing little astronomy and so gives only a superficial description of the achievements of the great Renaissance Danish astronomer Tycho Brahe (1546-1601). Written for the 400-year anniversary of Tycho's birth, the book is a tribute to him and social portrait of his times. Pleasant reading, but one must look elsewhere for full, critical explanations of Tycho's extraordinary planetary observations, solar and lunar theories, and wrong-headed revision of the Copernican theory. Photos and drawings; bibliography; footnotes; index. 209 pp.

Graham, Shirley. *Your Most Humble Servant.* New York: Julian Messner, 1949.

Biography. Graham offers an interpretation of the life of the African American astronomer and almanac writer Benjamin Banneker. Her attempt to portray him as a hero who rose above the social forces of his times is certainly admirable, but she uses novelistic techniques, such as dialogue and dramatically blocked-out scenes, to enhance the story's appeal but succeeds only in making her subject sound silly at times. She also quotes from Banneker's letters extensively. Bibliographic essay. 235 pp.

Hall, A. Rupert. *Isaac Newton, Adventurer in Thought.* Oxford, England: Blackwell, 1992.

Biography. To Hall, Isaac Newton had the greatest mind in English history. Although he credits Newton with profound insights into Christian history and prophesy, Hall concentrates on Newton's discoveries in mathematics, such as calculus, and physics, particularly the theory of gravitation and principles of optics, explicated for general readers who remember their algebra. Hall admits that Newton was touchy, intransigent, and cold in personal affairs but refuses to see Newton as a psychopathic genius—after all, Hall reasonably points out, Newton had many friends among scientists, if more enemies. A tone is reverential as the book provides a thorough review of Newton's best known intellectual achievements. Portrait of Newton; bibliography; end notes; index. 468 pp.

Haramundanis, Katherine, ed. *Cecilia Payne-Gaposchkin: An Autobiography and Other Recollections*. Cambridge, England: Cambridge University Press, 1984.
Autobiography-biography. Haramundanis, Jesse L. Greenstein, and Peggy A. Kidwell each provide introductions to *The Dyer's Hand* by Cecilia Payne-Gaposchkin (1900-1979): one emphasizes the content, the next discusses the historical context, and the last contains the editor's personal recollections of Payne-Gaposchkin. These take up about a quarter of the book. Then comes the autobiography, in which Payne-Gaposchkin, an able writer, recalls her youth, education, difficult tenure in the shadow of Harlow Shapley, work on stellar spectra and astrophysics, and recognition at Harvard University, where she was the first woman to become a full professor and to chair a department. A valuable resource on the history of women in American astronomy, the book is suitable for general readers, given Payne's skill at explaining difficult astrophysical principles. Photos; index. 269 pp.

Hindle, Brooke. *David Rittenhouse*. Princeton, N.J.: Princeton University Press, 1964.
Biography. To Hindle, the American natural philosopher David Rittenhouse (1732-1796) epitomizes the American enlightenment. Of humble origins, Rittenhouse showed an early talent for mathematics, which he put to good use as a surveyor and an astronomer measuring the transit of Mercury and Venus and observing comets and meteors. He also made precision clocks and orreries, spoke several language, participated in the first Constitutional Convention and the Revolution, was a member of the Royal Society, and became president of the American Philosophical Society. Among his friends and admirers were Thomas Jefferson, Benjamin Franklin, and Benjamin Rush. Photos and drawings; bibliographical essay; index. 394 pp.

Hoskin, Michael A. *William Herschel and the Construction of the Heavens*. New York: W. W. Norton, 1964.

Scientific biography. Hoskin focuses strictly on William Herschel's ideas about stars and their distribution in the heavens. Hoskin explicates Herschel's developing ideas by close analyses of the astronomer's most important papers published between 1783 and 1817. A useful resource for readers who already know a good deal about Herschel, geometry, and basic astronomy. Photos, drawings, and diagrams; bibliographical essay; end notes; index. 199 pp.

Hoyle, Fred. *Home Is Where the Wind Blows: Chapters from a Cosmologist's Life.* Mill Valley, Calif.: University Science Books, 1994.
Autobiography. Fred Hoyle (b. 1915) reminisces about his intellectual development, beginning with his youth in northern England and continuing through Cambridge University, service during World War II, and his career as an astrophysicist, cosmologist, and popular author. The story lies at the heart of cosmology through the third quarter of the twentieth century. Hoyle co-authored the steady-state cosmological model, long a rival of the big bang model, and helped develop the theory explaining the formation of heavy elements. He also headed the Institute for Theoretical Astronomy in Cambridge while serving as an astronomy professor there. His wide-ranging, usually controversial interests included human overpopulation, the origin of life, and Stonehenge. Photos; index. 443 pp.

Hoyle, Fred. *Nicolaus Copernicus.* New York: Harper and Row, 1973.
Biography. This highly readable little book by Fred Hoyle, an eminent astrophysicist and science popularizer, contains only a short sketch of the life of Nicolaus Copernicus, but it lays out his career clearly. The rest of the book is a close analysis of the Copernican model of the universe, and readers will need to know geometry and trigonometry to follow it. Drawings and diagrams; short bibliography; index. 94 pp.

Koestler, Arthur. *The Watershed: A Biography of Johannes Kepler.* Garden City, N.Y.: Anchor Books, 1960.
Biography. An excerpt from Koestler's *The Sleepwalkers: A*

History of Man's Changing Vision of the Universe. Koestler emphasizes the relation between Johannes Kepler's intellectual and psychological development, whose roots lay in the cosmologist's harsh, guilt-ridden youth. A major thesis of the book is that a false inspiration about the geometric structure of the universe, maintained with paranoid conviction and intensity, propelled Kepler to his greatest discovery, the laws of planetary motion, which inaugurated modern cosmology. Kepler's life is not a cheery story in the least, and Koestler capitalizes on that fact to create a morbidly fascinating portrait, as well as a very readable work on intellectual history. Drawings; end notes; index. 280 pp.

Lovell, Bernard. *Astronomer by Chance.* New York: Basic Books, 1990.
Autobiography. Founder and director of the Jodrell Bank radio telescope facility in England, Bernard Lovell (b. 1913) recalls his career as a pioneer radio astronomer. He also took part in developing radar for England's defense during World War II, and the episode is the most gripping section of the book. Most of his story, however, concerns the scientific discoveries made with the radio telescope and the administrative and financial problems of a large science project. Photos; glossary; index. 381 pp.

McPherson, Stephanie Sammartino. *Rooftop Astronomer: A Story About Maria Mitchell.* Minneapolis, Minn.: Carolrhoda Books, 1990.
Biography for readers ten years and older. McPherson concentrates on the circumstances that led Maria Mitchell (1818-1889) to become an astronomer. The book relates her childhood in Nantucket, teaching at Vassar College, discovery of a comet, and election as the first woman member of the American Association for the Advancement of Science. McPherson also depicts Mitchell's efforts to get more women to enter professions. Drawings; bibliography. 64 pp.

Meadows, A. J. *Science and Controversy: A Biography of Sir Norman Lockyer.* London: Macmillan, 1972.
Biography. This engagingly written biography not only traces

the varied career of Norman Lockyer (1836-1920); it also portrays the British scientific community in the late nineteenth century. Lockyer was a pioneer in solar spectroscopy and worked on topics in meteorology, archeology, and chemistry, discovering helium. His influence on science of the times was broad. A member of the Royal Society, he founded the science journal *Nature* and edited it for 50 years. According to Meadows, Lockyer tended to come out the worse in the controversies of the day, but the fact hardly discouraged this ebullient man. For general readers. Photos; bibliography; end notes; index. 331 pp.

Milne, Edward Arthur. *Sir James Jeans*. Cambridge, England: Cambridge University Press, 1952.
Scientific biography. This review of the life and ideas of the astrophysicist James Jeans (1877-1946) is by the equally famous cosmologist E. A. Milne. Not surprisingly, Milne concentrates on scientific topics: the state of science when Jeans was a boy, the partition of energy, rotating fluid masses (such as the Sun), star clusters, and the equilibrium of stars, all written for those who understand advanced mathematics and physics. But Milne also devotes long chapters to Jeans' youth, education, and university career, as well as his tenure as secretary of the Royal Society. Milne concludes with a discussion of Jeans's philosophy of science, which he finds disappointing. Photos; bibliography; index. 176 pp.

Newcomb, Simon. *The Reminiscences of an Astronomer*. Boston, Mass.: Houghton Mifflin, 1903.
Autobiography. During his long career, the American astronomer Simon Newcomb (1835-1909) specialized in mathematical calculations of planetary motion. He was especially well known for his calculations concerning the transit of Venus, the orbits of asteroids, and the motion of the Moon. His recollections also depict the scientific community and operations of government-supported scientific projects in nineteenth-century America. Photo of author; index. 424 pp.

Redondi, Pietro. *Galileo, Heretic*. Princeton, N.J.: Princeton University Press, 1987.

Biography-history. This book concerns only Galileo Galilei's trial for heresy, which the author re-examines in great depth, based in part on the discovery of a new document. Redondi concludes that the Catholic church's condemnation of Galileo for teaching Copernicanism was a subterfuge intended to hide the real trouble: Galileo's atomistic views endangered the orthodox view of the Eucharist. Photo reproductions of artwork and documents; footnotes; index. 356 pp.

Reston, James, Jr. *Galileo: A Life*. New York: HarperCollins, 1994.
Biography. A capably written book that concentrates on Galileo Galilei's struggles with academic, governmental, and Catholic politics. In particular, Reston argues that the church's success in forcing Galileo to recant his support for Copernican cosmology marked the beginning of the decline in Italy's cultural prestige and has been a black mark on church history that ecclesiastical leaders have still not successfully erased. Reston describes Galileo's major scientific discoveries in passing for general readers and devotes some space to his family life, especially his relationship with his elder daughter. Drawings; end notes; bibliography; index. 319 pp.

Ronan, Colin A. *Edmond Halley: Genius in Eclipse*. Garden City, N.Y.: Doubleday, 1969.
Biography. Ronan's thesis is that Edmond Halley worked in the shadow of the great genius Isaac Newton; nevertheless, Halley was a brilliantly original scientist in his own right. Ronan's engagingly written narrative carries home the point well, showing Halley's incredible industry and intellect in astronomy (achieving far more than calculating the orbit of Halley's Comet), physics, oceanography, biology, and technology. In fact, he epitomized the then novel philosophy of observation and experimentation. Furthermore, to Halley belongs much of the credit for publishing Newton's famed *Principia mathematica* in 1686. Halley, a man in love with ideas, is an attractive subject in Ronan's skillful treatment. Photos and drawings; bibliography; end notes; index. 251 pp.

Shapley, Harlow. *Through Rugged Ways to the Stars*. New York: Charles Scribner's Sons, 1969.

Autobiography. A key figure in astronomy during the first half of the twentieth century, Harlow Shapley helped measure the Milky Way and determine the solar system's position in it. These loose memoirs, based upon oral history transcripts, range over his education and career with some humor and generous acknowledgments of others. He downplays the importance of his "great debate" with Heber D. Curtis about the Milky Way, often touted as a turning point in modern astronomy. As director of the Harvard College Observatory, Shapley wielded considerable power in American science, and he describes his experiences with scientific politics, which included visits to Soviet astronomers and wild accusations that Shapley was a communist by Senator Joseph McCarthy. Photos; bibliography; index. 180 pp.

Sharov, Alexander S., and Igor D. Novikov. *Edwin Hubble, The Discoverer of the Big Bang Universe*. Cambridge, England: University of Cambridge Press, 1993.
Biography-popular science. The original edition of this book was intended to introduce the life and discoveries of Edwin P. Hubble to Russian readers. This readable translations is a succinct, lucid overview divided into two sections. The first is biographical, emphasizing the development of Hubble's interests, astronomical ideas, and career. The second part explains how Hubble's constant and the expansion of the universe have led to modern investigations into the large-scale structure of the universe and the big bang cosmological model. Readers who know basic astronomy will benefit best from the text. Photos and diagrams; bibliography; indexes. 187 pp.

Shklovsky, Iosif. *Five Billion Vodka Bottles to the Moon: Tales of a Soviet Scientist*. New York: W. W. Norton, 1991.
Autobiography. Shklovsky (1916-1985) was a leading Russian astronomer of the twentieth century and a pioneer in radio astronomy. This book, however, spends most of its time on the politics of science in the Soviet Union. The twenty-four self-contained stories involve a host of famous names, such as Andrei Sakharov and Carl Sagan, and incidents, such as the infamous "doctors plot" of 1953 and the first Soviet eclipse

expedition to Brazil in 1947. Shklovsky's humor, continuously displayed, is mordant and his opinions strong, making the stories enjoyable, provocative reading. Drawings and photos; index. 268 pp.

Sidgwick, J. B. *William Herschel: Explorer of the Heavens.* London: Faber and Faber, 1953.
Biography. Sidgwick devotes the narrative to William Herschel's astronomical discoveries, astronomy in general during the late eighteenth and early nineteenth centuries, and his design and construction of telescopes. Herschel's private life, although outlined, does not receive analytical attention; his sister, Caroline, and son, John, figure in the story, but almost wholly in their contributions to Herschel's work. His studies of sidereal motion, sky surveys, and theory of island universes occupy most of the text. Photos; short bibliography; index. 228 pp.

Szperkowicz, Jerzy. *Nicolaus Copernicus, 1473-1973.* Warsaw, Poland: Panstwowe Wydawn, 1973.
Biography. A pamphlet published to commemorate the 500-year anniversary of Nicolaus Copernicus' birth. Not surprisingly, it depicts him as a scientific hero. The explanation of the heliocentric model of the universe, while lucid and readable as far as it goes, is sketchy. Photos and drawings. 87 pp.

Thomas, Henry. *Copernicus.* New York: Julian Messner, 1960.
Biography for readers ten years and older. In this lengthy but simply written account, Thomas limns the cultural background and state of scientific knowledge during the era of Nicolaus Copernicus. But he reserves most of the text for explaining how Copernicus' heliocentric model of the universe departed from established belief and the angry reactions to it. Bibliography; index. 192 pp.

Thoren, Victor E., with John R. Christianson. *The Lord of Uraniborg: A Biography of Tycho Brahe.* New York: Cambridge University Press, 1990.
Biography. A well-written account of one of the great astronomers of the Renaissance, the Danish nobleman Tycho

Brahe. Thoren emphasizes the importance of Brahe's aristocratic rank and privilege as preparation for his astronomical work, much of it done from an observatory Brahe built on the island Uraniborg. His many achievements include the discovery of a supernova and a wrong-headed reinterpretation of the Copernican theory, but he is best known for the meticulous planetary observations that allowed his assistant, Johannes Kepler, to formulate laws of planetary motion. Although highly technical in some chapters, the book should appeal to all readers interested in science history or in a tale of a romantic genius who fought duels, quarreled incessantly, invented his own instruments, and explored the unknown. Drawings and diagrams; footnotes; indexes. 523 pp.

Wali, Kameshwar C. *Chandra: A Biography of S. Chandrasekhar.* Chicago, Ill.: University of Chicago Press, 1991.
Biography. Subrahmanyan Chandrasekhar (1910-1994) was a leading figure in twentieth-century astrophysics. While still a teenager, he discovered a fundamental limitation to the mass of white dwarfs, a finding so unexpected that leading astrophysicists, some his teachers, rejected it. Wali tells Chandrasekhar's story well for non-scientists: his youth in a progressive Indian family, education at Cambridge University, immigration to the United States, professorship at the University of Chicago, editorship of the prestigious *Astrophysical Journal*, and influence on hundreds of students. The biography, in fact, concentrates on his personal life and career; Wali describes his discoveries in physics and astrophysics non-technically. An engrossing book that reveals how a shy, supremely disciplined young genius crossed personal and cultural boundaries to rise to the top of his abstruse profession. Photos; appendix containing a long interview with Chandrasekhar; end notes; index. 341 pp.

Webb, George Ernest. *Telescopes and Tree Rings.* Tucson: University of Arizona Press, 1983.
Biography. Webb's subject is Andrew Ellicott Douglass (1867-1962), who served as Percival Lowell's chief assistant at Lowell Observatory until the two fell out over how to interpret observations of canal-like markings on Mars. Douglass

went on to found Steward Observatory in Arizona. There he became interested in tree rings, especially the information they can provide about fluctuations in Earth's climate. Photos; index; bibliography; end notes. 242 pp.

Wright, Helen. *Explorer of the Universe: A Biography of George Ellery Hale*. New York: E. P. Dutton, 1966.
Biography. George Ellery Hale (1868-1938) was an energetic man. He planned and helped build three great observatories: the Yerkes, Mount Wilson, and Palomar. Palomar for a long time housed the world's largest telescope, named after Hale. As Wright relates, the discoveries made at these observatories revolutionized astronomy and cosmology. Hale was an astronomer in his own right, a specialist in solar astronomy and pioneer in solar spectroscopy. Yet, according to Wright, his influence went well beyond astronomy; Hale was in fact instrumental in fostering American science and the humanities in the early twentieth century. She carefully shows how he developed from a young boy interested in science to an administrator and leader who dealt with powerful men of the times, such as Andrew Carnegie. For general readers. Photos; bibliography; end notes; index. 480 pp.

Wright, Helen, Joan N. Warnow, and Charles Weiner, eds. *The Legacy of George Ellery Hale*. Cambridge, Mass.: MIT Press, 1972.
Biography-history. This is primarily a picture book, but the text includes a short biography of George Ellery Hale (by Wright), reprints of five of his articles, and articles by others concerning Hale's influence on American astronomy and science administration. Many photos, drawings, reproductions of documents and publications, cartoons, and diagrams; end notes; index. 293 pp.

Zinner, Ernst. *Regiomontanus: His Life and Work*. Amsterdam, The Netherlands: North-Holland, 1990.
Biography. Zinner describes the life and times of the wandering astronomer and mathematician Johannes Müller (1436-1476), later dubbed Regiomontanus. A brilliant man who mysteriously disappeared, Regiomontanus made astronomi-

cal instruments for his own observations, wrote almanacs, and promoted calendar reform, but the book also portrays the pre-scientific tenor of the age, especially its interest in astrology. A fascinating life; unfortunately, this book is strictly for scholars and dedicated readers who know the history of mathematics, Latin, and German. A supplement contains articles by other writers about Regiomontanus' life and mathematics. Bibliography; end notes; index. 402 pp.

Chapter 3

Chemistry

COLLECTIONS

The Biographical Dictionary of Scientists: Chemists. New York: Peter Bedrick Books, 1984.

Following a brief history of chemistry, this volume has articles on 169 chemists, as early as Robert Boyle (1627-1691). Averaging about five hundred words in length, entries include biographical information and summaries of the scientists' discoveries for general readers, although there is some use of chemical notation. A handy basic reference because of the extent of coverage and the brevity of the articles. Drawings; glossary; index. 203 pp.

Davis, Kenneth S. *The Cautionary Scientist: Priestley, Lavoisier, and the Founding of Modern Chemistry.* New York: G. P. Putnam's Sons, 1966.

Biography-history. As the title suggests, this book has a polemical thesis, that in abandoning the why of things and concentrating instead on the what and how of things, modern thinkers have abdicated moral responsibility for their inventions. Davis traces the origin of "progress" as an informing cultural principle to the Age of Reason, and he discusses the lives and careers of the English Protestant theologian-scientist Joseph Priestley and the Frenchman Antoine Lavoisier as examples of men who balanced the why and the how in their scientific work, as, according to the author, overspecialized modern scientists no longer do. The text explains their scientific ideas for general readers, and the emphasis is upon philosophy as much as science. Diagrams; end notes; bibliography; index. 256 pp.

Farber, Eduard. *Nobel Prize Winners in Chemistry*. New York: Henry Schuman, 1953.
Farber covers prizes awarded from 1901 to 1950, fifty-three laureates altogether. A short biographical sketch opens each entry, followed by an excerpt of about five hundred words from the laureate's publications or the Nobel lecture and a shorter comment on the consequences of the laureate's achievement in chemical theory and practice. The aim is to provide the public with more accurate information about Nobel winners than the press provides yearly when the winners are announced. Photos; indexes. 219 pp.

Findlay, Alexander, and William Hobson Mills, eds. *British Chemists*. London: The Chemical Society, 1947.
Written for scientists, this book describes the lives and achievements of sixteen chemists in the nineteenth and twentieth centuries: William Crookes, James Dewar, Henry Edward Armstrong, Raphael Meldola, Harold Baily Dixon, William Ramsay, William Henry Perkin, Arthur George Perkin, Arthur Green, Arthur Harden, William Pope, Gilbert Morgan, Arthur Lapworth, Jocelyn Field Thorpe, Thomas Lowry, and George Barger. A useful resource for historians of chemistry because the entries reveal the esteem the scientists enjoyed from their colleagues. Photos; extensive use of chemical notation and in-text source citation. 431 pp.

Harrow, Benjamin. *Eminent Chemists of Our Time*. 2d ed. New York: D. Van Nostrand, 1927.
The "our time" of the title is the late nineteenth and early twentieth centuries. Harrow devotes a chapter each to the lives and works of 11 chemists: William Henry Perkin, Dmitry Ivanovitch Mendeleyev, William Ramsay, Theodore Richards, Jacobus Van't Hoff, Svante Arrhenius, Henry Moissan, Marie Curie, Victor Meyer, Ira Remsen, and Emil Fischer. Written for general readers, the biographical section of each chapter emphasizes the scientist's personal life; an accompanying essay covers the science, and Harrow labors to show a devel-

opment of the discipline through a foundation period, a classification period, a physico-chemical period, and a radioactivity period. The book is full of biographical information not easily found elsewhere, especially for figures who are no longer well known. Photos; bibliography with each essay; index. 471 pp.

Jaffe, Bernard. *Crucibles: The Story of Chemistry*. New York: Simon and Schuster, 1948.
Biography-history. Jaffe recounts the history of chemistry through portraits of 17 scientists and a separate essay discussing those who discovered nuclear fission. Arranged chronologically, the book begins with Bernard Trevisan (1406-1490) and ends with Ernest O. Lawrence (1901-1958). Jaffe writes for general readers so that they can better appreciate both the promise and threat of modern chemistry, especially in relation to nuclear energy. The text is dramatic and intended to entertain, occasionally featuring imagined dialogue and fanciful descriptions as well as historical information. It is in fact enjoyable reading, although because of the author's enthusiasm, the text should be treated with caution. End notes; bibliography; index. 480 pp.

James, Laylin K., ed. *Nobel Laureates in Chemistry, 1901-1992*. Washington, D.C.: American Chemical Society and the Chemical Heritage Foundation, 1993.
Written to bring new perspectives on chemistry to a wider audience, this volume examines the scientific achievements of 116 laureates. Each entry is written by a university scholar. The essays, about 2,000 to 3,000 words long, allot little space to personal biography; the focus is the laureate's career, with special attention to the discovery that earned the prize. Readers must be familiar with the technical vocabulary of chemistry and its basic principles to fully benefit from the scientific explanations. A select bibliography of the laureate's publications concludes each entry. Photos of laureates; general bibliography; footnotes; index. 798 pp.

Nobel Prize Winners: Chemistry. 3 vols. Pasadena, Calif.: Salem Press, 1990.

Part of a series on Nobel Prize winners, this set offers sophisticated explanations of the work of 113 winners from 1901 to 1989. Scientists and science historians wrote the entries, which follow a standard format: a description of the award ceremony; a summary of the Nobel lecture; a summary of the critical reception of the award in the scientific community and popular press; a biographical sketch; an extended consideration of the laureate's entire scientific career; and an annotated bibliography—about 3,500 words in all. The sections on science address topics in depth for educated general readers. A much-praised reference work with reliable information and a readable presentation. Photos; indexes.

Miles, Wyndham D., ed. *American Chemists and Chemical Engineers*. Washington, D.C.: American Chemical Society, 1976.
The editor wants his collection to preserve knowledge about a "small selection of American chemists who ought not to be forgotten." From the alchemists of colonial times, about 300 years ago, to chemists who died just before the book's publication, the editor chose approximately five hundred scientists, administrators, and editors important to the development of chemistry and its technology. The entries, written by scientists and historians, run from 100 to 1,000 words and cover in about equal proportions the subjects' personal lives, careers, and work in chemistry. A short bibliography ends each entry. Index. 544 pp.

Smith, Henry Monmouth. *Torchbearers of Chemistry*. New York: Academic Press, 1949.
This book is primarily a portrait gallery. Arranged alphabetically, more than 250 photographs, reprints of paintings and engravings, and caricatures let readers gaze at the features of famous chemists, natural philosophers, and physicists. The coverage is international and extends from Jabir Ibn Hayyan (ca. 720-813) to Moses Gomberg (1866-1947). A brief paragraph concerning the life and achievements of each scientist accompanies the portrait. A bibliography of biographies of the scientists concludes the volume and is its most apposite feature for scholars. 270 pp.

BIOGRAPHIES AND AUTOBIOGRAPHIES

Barton, Derek H. R. *Some Recollections of Gap Jumping.* Washington, D.C.: American Chemical Society, 1991.
Scientific autobiography. Derek H. R. Barton (b. 1918) won the Nobel Prize in Chemistry in 1969 for his work on conformational analysis. During his long career he taught at Imperial College, Harvard University, Texas A&M University, and Birbeck College and served as director of research at the Centre National de la Recherche Scientifique in France. Barton dwells on his scientific work; he is rather casual in presenting the times and dates of his career and comments little about his personal life, although he does reminisce about colleagues. The scientific passages rely on chemical notation. Photos; end notes; index. 143 pp.

Berry, A. J. *Henry Cavendish: His Life and Scientific Work.* London: Hutchinson, 1960.
Biography. Berry sums up the "uneventful" life of Henry Cavendish (1731-1810) in the opening chapter. Thereafter, he devotes chapters to the main subjects of Cavendish's physical and chemical research. These included hydrogen and other gases, electricity, combustion and heat, the chemical composition of water (for which he is best known), and nitric acid. Cavendish was among the first scientists to use calculus to analyze physical phenomena and also constructed meteorological and astronomical instruments. Berry concludes with an appraisal of Cavendish's influence on modern science. A scholarly biography but accessible to lay readers interested in the history of science. Some chemical notation is used. Photos and diagrams; glossary; footnotes; short bibliography; index. 208 pp.

Berzelius, Jöns Jacob. *Autobiographical Notes.* Edited by H. G. Söderbaum. Baltimore, Md.: Williams and Wilkins, 1934.
Autobiography. The editor constructed this book from autobiographical sketches that Jöns Jacob Berzelius (1779-1848) submitted to the Swedish Academy of Sciences as a condition of membership. An early champion of the atomic hypothesis, Berzelius sought to classify minerals on a chemical basis, iso-

lated and named thorium (after the Norse god Thor), and advanced an electrochemical theory based upon his own experiments. In this book he recollects key stages of his career and education. Portrait of Berzelius; end notes. 194 pp.

Birch, Beverly. *Louis Pasteur*. Milwaukee, Wis.: Gareth Stevens, 1989.
Biography for readers 12 years and older. Supported by many eye-catching illustrations, Birch emphasizes Louis Pasteur's research in establishing the germ theory of diseases. She opens the book with a dramatic account of Pasteur's trip high into the Alps to collect air samples, but the reader does not find out what the samples were for (to prove microbes are everywhere) until Birch backtracks to explain the controversy over the germ theory and Pasteur's background, education, and career. She also discusses his research in crystals and crop diseases. Photos, artwork, and drawings; bibliography; glossary; index. 68 pp.

Calvin, Melvin. *Following the Trail of Light: A Scientific Odyssey*. Washington, D.C.: American Chemical Society, 1992.
Scientific autobiography. Melvin Calvin (b. 1911) studied a wide variety of topics in organic chemistry and is best known for his research on photosynthesis, for which he won the 1961 Nobel Prize in Chemistry. After three modest chapters on his family background, education, and university career, he discusses the technicalities of his research. Among the topics are the theory of organic chemistry, photochemistry, organic crystals, free radicals, chemical and viral carcinogenesis, and hydrocarbons. He also recounts the building of the Laboratory of Chemical Biodynamics at the University of California, Berkeley, where he was a professor, and his work during World War II. The many passages about chemistry require an understanding of its basic ideas and notation. Photographs and diagrams; end notes; index. 175 pp.

Cameron, Frank. *Cottrell, Samaritan of Science*. Garden City, N. Y.: Doubleday, 1952.
Biography. Frederick Gardner Cottrell (1877-1948) was an industrial scientist specializing in physical chemistry. He

invented devices for precipitating suspended particles and studied nitrogen fixation and sea water conversion. He rose to become director of the U.S. Bureau of Mines, director of the American Coal Company, and chairman of the National Research Council; such positions gave him great influence in science during the early twentieth century: for example, he helped arrange crucial grants for Ernest O. Lawrence's cyclotron at the Radiation Laboratory in Berkeley, California. According to Cameron, Cottrell was a man of wide interests, great ability, and abundant goodwill, which this richly detailed biography is intended to reveal to general readers. Index. 414 pp.

Conner, Edwina. *Marie Curie*. New York: Bookwright Press, 1987. Biography for readers eight years and older. Conner emphasizes both the dangers and the intellectual triumphs of Marie Curie's career. Through incredible devotion to her studies, she left Poland, attended university in Paris, worked with her husband, Pierre, to isolate radium, provided medical care to the wounded during World War I, was showered with honors, including two Nobel Prizes, and died of radiation poisoning from her work. Conner presents the story clearly and does a good job explaining the basic nature of the Curies' work in physics and chemistry. Photos and drawings; short bibliography; glossary; index. 32 pp.

Cram, Donald J. *From Design to Discovery*. Washington, D.C.: American Chemical Society, 1990.
Scientific autobiography. After a brief review of his upbringing and education, Donald J. Cram (b. 1919) launches into a series of sophisticated reviews of his research interests. These included phenonium ions, the stereochemistry of carbanions and carbonyl addition reactions, and crown ethers. His research on the last earned him the Nobel Prize in Chemistry for 1987. A professor at the University of California at Los Angeles, Cram also wrote an influential chemistry textbook. Readers need a sophisticated understanding of chemistry and its notation to appreciate the text. Photos and diagrams; end notes; index. 146 pp.

Crosland, Maurice. *Gay-Lussac, Scientist and Bourgeois*. Cambridge, England: Cambridge University Press, 1978.
Biography. Among France's greatest scientists during the nineteenth century, Joseph-Louis Gay-Lussac (1778-1850) attended the École Polytechnique during the turbulent early Napoleonic era, studied the expansion of gases, ascended in a primitive hot-air balloon, conducted research on capillarity and various types of chemical synthesis, analyzed organic compounds, wrote about absolute zero, became a professor at his alma mater, and was elected to the Chamber of Deputies during the reign of Louis Philippe. To Crosland, Gay-Lussac epitomizes the new breed of scientist to emerge from post-revolutionary France, the first to take science as a full-time profession. He also is modern in that he participated in many joint publications of scientific papers, a practice almost unknown before 1800. Crosland tells Gay-Lussac's story to illustrate the rise of the modern scientist. For general readers interested in the history of science. Portrait of Gay-Lussac and diagrams; bibliography; end notes; indexes. 333 pp.

Curie, Eve. *Madame Curie, a Biography*. New York: Doubleday, Doren, 1938.
Biography. The author combines her own memories of her mother, Marie Curie (1867-1934), with those of her family, Marie's friends, and official documents. The result is an intimate view of a person who "did not know how to be famous." Marie Curie, to her daughter, remained devoted to science, uncorrupted by fame, her whole life, giving freely of her energy to help the sick and wounded during World War I and to teach others. The book contains only general accounts of Curie's discoveries in chemistry and physics (she won Nobel Prizes in both disciplines), and the author often quotes from her mother's papers about technical matters. One photo; index. 393 pp.

Davis, Joseph S., ed. *Carl Alsberg, Scientist at Large*. Stanford, Calif.: Stanford University Press, 1948.
Biography. The five essays in this volume eulogize Carl Alsberg (1877-1940), an influential agricultural chemist. The essays address his life, work on seafood and poisons, tenure

as chief of the U.S. Bureau of Chemistry, and university career. Three essays by Alsberg on chemistry and its relation to other natural and social sciences conclude the book. For readers with a basic knowledge of chemistry. Photo of Alsberg; bibliography; index. 182 pp.

Dewar, Michael J. S. *Semiempirical Life.* Washington, D.C.: American Chemical Society, 1992.
Scientific autobiography. Michael J. S. Dewar was born in India in 1918 to parents in the Civil Service; he attended Oxford, where, he says, he was largely self-taught, he and worked in both industrial and academic chemistry in England and the United States. An unusual life and a charming, often funny book, despite the complexity of the science discussed. As well being as a prominent theoretician, Dewar gained attention with his studies of boron chemistry, superconductors, the biosynthesis of fatty acids, and phenyl radicals. Like many scientists, he acknowledges that his scientific career sprang indirectly from his youthful interest in science fiction. Although Dewar does not employ chemical notation extensively, the text requires familiarity with advanced chemistry. Photos and diagrams; end notes; index. 215 pp.

Djerassi, Carl. *Steroids Made It Possible.* Washington, D.C.: American Chemical Society, 1990.
Scientific autobiography. Carl Djerassi (b. 1923) pioneered synthesis of steroids as an industrial chemist and invented the first oral contraceptive, for which he is best known. He also taught at universities, including Stanford, started his own artists' colony, and wrote fiction. Such a variety of undertakings, conveyed in a pugnacious and sometimes witty style, makes his book unusually absorbing reading for a chemist's autobiography. There is also a good deal of technical explanation, often based upon chemical notation, which only chemists will appreciate fully. Photos; end notes; index. 205 pp.

Eliel, Ernest L. *From Cologne to Chapel Hill.* Washington, D.C.: American Chemical Society, 1990.
Scientific autobiography. German-born Ernest L. Eliel (b. 1921) wrote an influential textbook on organic chemistry,

Stereochemistry of Carbon Compounds, conducted research in stereochemistry and conformational analysis, taught at Notre Dame and the University of North Carolina, and became chairman of the board of the American Chemical Society. Eliel recounts his childhood; years as a refugee from the Nazi regime in his homeland; studies in Europe, Cuba and the United States; and academic career and research. The passages about the research include chemical notation. Photos; end notes; index. 138 pp.

Fisher, Leonard Everett. *Marie Curie*. New York: Macmillan, 1994.
Biography for readers eight years and older. Fisher's dark, brooding black-and-white illustrations dominate the book and the well-written text, which recounts Marie Curie's development as a scientist and work with her husband, Pierre, to isolate radium. Fisher also describes her work in French battlefields during World War I to provide X-rays and radium therapy to the wounded. Drawings throughout. 29 pp.

French, Sidney J. *Torch and Crucible: The Life and Death of Antoine Lavoisier*. Princeton, N.J.: Princeton University Press, 1941.
Biography. Calling Antoine Lavoisier (1743-1794) one of the most versatile men of all time, French attempts to display the two sides of the man, his scientific research and his political role in revolutionary France. French carefully lays out the family history and rearing of Lavoisier before tracing his interest in geology and his education. French devotes the main share of the book to Lavoisier's theory of atomic chemistry, his discovery of the composition of water, and the controversies these achievements started. Then the book recounts Lavoisier's public service and execution by revolutionaries. The book is dramatically written, intended for all readers. Bibliography; index. 285 pp.

Geison, Gerald L. *The Private Science of Louis Pasteur*. Princeton, N.J.: Princeton University Press, 1995.
Scientific biography. Geison provides summary biographical information about Louis Pasteur (1822-1895) in preparation for his subject: the hidden side of Pasteur's scientific work.

Drawing from Pasteur's laboratory notebooks, as well as his published papers, Geison finds that the great French scientist involved himself in some shady, unethical methods and was not above plagiarizing to stay ahead of a rival. The book is a highly focused, closely analytical work, and a valuable resource to readers interested in the politics of science. Geison explains the chemical and biological ideas involved clearly and writes engagingly, but general readers should look to other sources for a full discussion of Pasteur's life and works. Photos, drawings, and diagrams; bibliography; end notes; index. 378 pp.

Getman, Frederick H. *The Life of Ira Remsen*. Easton, Pa.: Journal of Chemical Education, 1940.
Biography. A chemist who studied ozone and acids, Ira Remsen (1846-1927) was best known as an educator who put chemistry instruction in the United States on a firm foundation. He taught at Williams College and Johns Hopkins University (where he also served as president) and wrote eight textbooks. He also chaired a presidential panel investigating the purity of foods. Getman lauds the work of Remsen and quotes frequently from his correspondence. Photos; bibliography. 172 pp.

Gibbs, F. W. *Joseph Priestley: Revolutions of the Eighteenth Century*. Garden City, N.Y.: Doubleday, 1967.
Biography. Gibbs relates the life of Joseph Priestley (1733-1804) in broad outlines. A theologian and philosopher as well as a scientist, Priestley was both feared and admired by contemporary Englishmen depending upon their political sentiments, for Priestley was a forceful radical. Gibbs pauses in the story of Priestley's life to describe his scientific studies—electricity, gases, acids and alkalis, and gunpowder—in separate chapters. Pleasant reading for anyone interested in the history of science or life in eighteenth-century England and the United States. Portraits, drawings, and diagrams; bibliographical essay; index. 258 pp.

Giroud, Françoise. *Marie Curie, a Life*. New York: Holmes and Meier, 1986.

Biography. Giroud seized upon a comment by Albert Einstein to guide her interpretation of Marie Curie's life—that she was "the only person to be uncorrupted by fame." What temperament made that purity possible? Giroud expresses surprise for what she found. In this pointedly non-academic book, she explains why. Giroud focuses on Curie's rearing, education, and personal life, although her career receives extensive treatment and her scientific accomplishments are described non-technically. Photos; short bibliography. 291 pp.

Goertzel, Ted, and Ben Goertzel. *Linus Pauling: A Life in Science and Politics.* New York: Basic Books, 1995.
Biography. Linus Pauling was the first person to win two unshared Nobel Prizes, one in chemistry for research in the chemical bond and its role in the structure of complex molecules and a peace prize for his opposition to nuclear weapons and war in general. The first prize was not controversial, and the authors recount Pauling's scientific education and research in moderately non-technical detail. The second prize was extremely controversial, and the authors carefully explain the nature of the controversy and Pauling's battles with the U.S. government. They also discuss his well-known crusade late in life: the curative power of vitamin C. As a guiding theme, they ponder whether Pauling should be regarded as a "universal living hero." Their sporadic psychological analyses of him do not make him sound like an unqualified hero. Photos and diagrams; end notes; bibliography; index. 300 pp.

Goran, Morris. *The Story of Fritz Haber.* Norman: University of Oklahoma Press, 1967.
Biography. Fritz Haber (1868-1934) won the 1918 Nobel Prize in Chemistry for discovering nitrogen fixation, a great boon to the manufacture of artificial fertilizers. By that time, he was notorious among fellow scientists in his homeland, Germany, for his work on chemical weapons during World War I, which he undertook out of patriotism. (His service to his country notwithstanding, he had to flee Germany in 1933 to avoid Nazi persecution, because he was Jewish.) His career illustrates both the dangers and benefits afforded by scientific

research. Goran argues that Haber did not deserve contempt from other scientists for his weapons work. The explanations of chemical processes and their applications are generally clear, but the biographical information is handled clumsily. Photos; bibliography; index. 212 pp.

Graham, Jenny. *Revolutionary in Exile: The Emigration of Joseph Priestley to America 1794-1804*. Philadelphia, Pa.: American Philosophical Society, 1995.
Biography-history. Joseph Priestley disagreed with the British government's policy toward its North American colonies, took the American side during the Revolution, and lived the last ten years of his life in the United States. Graham examines the forces that made him leave England and his activities in the United States. Written for historians, the book concerns his philosophy, not his science. Portraits; bibliography; footnotes; index. 213 pp.

Guerlac, Henry. *Lavoisier—The Crucial Year*. Ithaca, N.Y.: Cornell University Press, 1961.
Biography-history. Guerlac closely examines the background and details of Antoine Lavoisier's 1772 experiment with combustion. The experiment, says Guerlac, marked the end of the phlogiston theory of combustion and the beginning of the chemical revolution that installed the atomic theory of chemistry. Although written for scholars, the book is accessible to lay readers interested in the history of science. Portraits; index. 240 pp.

Hager, Thomas. *Force of Nature: The Life of Linus Pauling*. New York: Simon and Schuster, 1995.
Biography. Hager tries to reveal the complexities in the personality of Linus Pauling (1901-1994), widely regarded as one of the twentieth century's greatest chemists. Certainly, as Hager shows, his scientific career was complex: Pauling performed significant work on chemical bonds, molecular structure, proteins, X-ray crystallography, sickle-cell anemia, quantum mechanics, immunology, and nutrition, including his famous crusade for the benefits of vitamin C. He won the Nobel Prize in Chemistry in 1954 and another for peace in

1962. Hager also discusses Pauling's political activism in depth and shows that as well as being capable of compassion and charm, Pauling was fiercely competitive and "emotionally constricted." A well-written study of both Pauling's science and his character, based in part on interviews with him. Photos; end notes; long bibliography; index. 721 pp.

Hahn, Otto. *Otto Hahn: A Scientific Autobiography*. New York: Charles Scribner's Sons, 1966.
Autobiography. A controversial figure, Otto Hahn (1879-1968) has a secure place in the history of twentieth-century science because he discovered uranium fission, for which he won the Nobel Prize in Chemistry for 1944. However, that discovery is the source of two large controversies: first, some accuse him of grabbing credit for the discovery from Lise Meitner, and second, he helped launch the complex modern era of nuclear bombs and atomic power. Hahn attributes to Meitner the correct interpretation of his experimental discovery in this book during his discussion of fission. He also relates his youth and student years, fellowships with William Ramsay and Ernest Rutherford, and his work on natural and artificial radioactivity at the Kaiser Wilhelm Institute for Chemistry. The main narrative is brief and mostly devoted to elucidating his research for college-educated general readers and his friendship with the leading physicists and chemists of the pre-World War II era. Substantial appendices contain crucial scientific papers by him. Photos and diagrams; section of biographical notes about contemporary scientists; bibliography; index. 296 pp.

Hartley, Harold. *Humphre Davy*. London: Nelson, 1966.
Biography. A man capable of flashes of genius transcending his contemporaries, Humphry Davy (1778-1829) did not really fulfill his potential, according to the author, despite great achievements and fame. Hartley focuses on Davy's science rather than his personal life, beginning the book with a chapter summarizing chemical knowledge at the end of the eighteenth century, when Davy became intellectually mature. Hartley reviews Davy's first experiments during boyhood,

tutelage by Thomas Beddoes at Bristol's Pneumatic Institute, study of nitrous oxide, experiments with electricity, lectures at the Royal Institution and elsewhere, discovery of potassium and sodium, isolation of alkaline metals, chemical philosophy, presidency of the Royal Society, and invention of a safety lamp. But Davy's marriage and travels also receive some attention. Photos, drawings, and diagrams; bibliography; index. 160 pp.

Havinga, Egbert. *Enjoying Organic Chemistry, 1927-1987.* Washington, D.C.: American Chemical Society, 1991.
Scientific autobiography. A technical account of the research of Leiden University chemistry professor Egbert Havinga (1909-1988). He investigated conformational analysis, photochemistry, aromatic photosubstitution, vitamins, peptides, and proteins. He was also admired as a teacher and supporter of graduate students. Photos and diagrams; end notes; index. 122 pp.

Holmes, Samuel Jackson. *Louis Pasteur.* New York: Harcourt, Brace, 1924.
Biography. Although best known as a pioneer microbiologist, Louis Pasteur (1822-1895) considered himself a chemist, as does Holmes in this uncritically laudatory book. Holmes passes quickly over the details of Pasteur's personal life, concentrating on the scientific disputes and, especially, the key experiments in his long career. The book is most valuable for its explications of Pasteur's work on molecular asymmetry, fermentation, diseases of wine and beer, silkworms, contagious disease (including anthrax and rabies) and the germ theory of disease, and vaccines. As Holmes tells the story, Pasteur does indeed seem a scientific hero, the rescuer of France's economy. Drawings and diagrams; index. 149 pp.

Huisgen, Rolf. *The Adventure Playground of Mechanisms and Novel Reactions.* Washington, D.C.: American Chemical Society, 1994.
Scientific autobiography. One of the most successful teachers of chemistry and most often cited researchers in Germany, Rolf Huisgen had a long career at the University of Munich

after World War II. A very brief first chapter of this book outlines his career, and then he launches into extensive technical discussion of his research in ring effects, benzyne chemistry, and electrolytic reactions, among other topics—discussions intended for chemists. A final chapter recounts his boyhood, university education, and family life and includes reflections on education, chemistry in general, and colleagues. Photos and drawings; end notes; index. 279 pp.

Knight, David. *Humphry Davy: Science and Power.* Oxford, England: Blackwell, 1992.
Biography. To the author, Humphry Davy was the "Newton of chemistry." Accordingly, Knight summarizes Davy's varied scientific work, emphasizing his romantic view of nature: experiments in electricity, isolation of the chemical elements sodium and potassium, and invention of a safety lamp. But Davy also was a great popularizer of chemistry, giving public lectures at the Royal Institution, and he rose to the presidency of the Royal Society. Thus, Knight tells us, he accrued power in government circles and over popular ideas, a scientific role that emerged in England in the early nineteenth century and became integral to modern society. In this sense, Davy was a precursor to modern scientists, but not in every way: He was also at home with poets and artists, writing poetry himself. A well-written account for educated general readers. Bibliography; end notes; index. 218 pp.

McKie, Douglas. *Antoine Lavoisier.* New York: Henry Schuman, 1952.
Biography. According to McKie, the appearance of Antoine Lavoisier's *Traité Élémentaire de Chimie* (*Elements of Chemistry*, 1789) revolutionized chemistry. He dispensed with the ancient four elements—air, fire, water, earth—and proposed that matter consists of compounds of the elementary substances. His work on water, combustion, and individual elements led him to this view, as MacKie explains for a general audience. He also tells of Lavoisier's political work as a social and economic reformer and his execution during the Revolution. Drawings and diagrams; bibliography; index. 440 pp.

Mark, Herman F. *From Small Organic Molecules to Large: A Century of Progress*. Washington, D.C.: American Chemical Society, 1993.
Scientific autobiography. Herman Marks (1895-1991) tells how he became one of the first scientists to apply new physical principles to chemistry in the early twentieth century. His work with electron defraction of gaseous molecules and polymers earned him a solid reputation, and among his students at the University of Vienna were Edward Teller and Leo Szilard before Mark fled Nazi antisemitism and came to the United States, where he worked at the Polymer Research Institute. Although he describes a good deal of chemistry in this book, it is not so technical as to exclude educated general readers, who may enjoy Mark's chatty passages about his many acquaintances in European and American science. Photos; short bibliography; end notes; index. 148 pp.

Melhado, Evan M., and Tore Frängsmyr, eds. *Enlightenment Science in the Romantic Era: The Chemistry of Berzelius and Its Cultural Setting*. New York: Cambridge University Press, 1992.
Scientific biography. The first essay of this volume reviews the life of Jöns Jacob Berzelius and prevailing ideas of his era. The next seven essays take up aspects of his chemical and geological research in detail. The final essays describes his travels in Europe. Scholars of the history of science and intellectual history are the primary audience for this book. A photograph and drawings; bibliography; footnotes; index. 246 pp.

Mendelssohn, Kurt. *The World of Walther Nernst*. Pittsburgh, Pa.: University of Pittsburgh Press, 1973.
Biography-history. The author writes about his mentor, Walther Nernst (1864-1941), who won the 1920 Nobel Prize for Chemistry for formulating the third law of thermodynamics. But the book is as much about German history and culture as about Nernst, and it is never quite clear whether the reader is to consider Nernst the best of that culture, a symbol of it, or simply the beneficiary of it. The author also casts himself in the book as a survivor returning to the scenes of his youth. In any case, Nernst rose to the top of German science and was a firm patriot, helping to introduce gas warfare in World War I

in order to bring Germany victory. The Nazis, however, undid all that Nernst stood for, and he died a disappointed man. A confusing but immensely informative book for general readers with a keen interest in history. Photos; index. 191 pp.

Merrifield, Bruce. *Life During a Golden Age of Peptide Chemistry: The Concept and Development of Solid-Phase Peptide Synthesis.* Washington, D.C.: American Chemical Society, 1993.
Scientific autobiography. As the title makes clear, Bruce Merrifield (b. 1921) tells how he devised a way to synthesize peptides. His work greatly invigorated peptide chemistry and earned him the Nobel Prize for Chemistry in 1984. Merrifield also relates the origin of his interest in chemistry and speaks of his large family and career at Rockefeller University. The explanations of his research require advanced training in chemistry to understand. Photos and diagrams; end notes; index. 297 pp.

Morselli, Mario. *Amedeo Avogadro, a Scientific Biography.* Dordrect, Netherlands: D. Reidel Publishing, 1984.
Biography. Amedeo Avogadro (1776-1856) advanced the molecular hypothesis of chemistry. Virtually ignored during his lifetime, the hypothesis became the vital center of modern chemistry. Trained in jurisprudence, he was largely self-taught as a scientist. He worked outside the mainstream of scientific thought. A theoretician rather than experimenter, Avogadro approached chemistry as a part of physics. Morselli explains Avogadro's ideas in detail and fits them into their historical milieu. While the text is intellectually sophisticated, it is not beyond readers who understand basic science. Diagrams; bibliography; end notes; indexes. 375 pp.

Nakanishi, Koji. *A Wandering Natural Products Chemist.* Washington, D.C.: American Chemical Society, 1991.
Scientific autobiography. Koji Nakanishi (b. 1925) came to the United States from his birthplace, Hong Kong, via Lyon, France, London, Alexandria, Egypt, and wartime Japan. The story of his youth is fascinating and engagingly told. Unfortunately for lay readers it takes up only the first small sections of the book. Thereafter, Nakanishi relates his aca-

demic and industrial careers in Japan and the United States (Columbia University) and his chemical research. Although he speaks of such famous colleagues and friends as Carl Djerassi and of his interest in magic tricks, he devotes much of the text to topics only a chemist will appreciate fully. Photos and diagrams; end notes; index. 230 pp.

Nozoe, Tetsuo. *Seventy Years in Organic Chemistry*. Washington, D.C.: American Chemical Society, 1991.
Scientific autobiography. Tetsuo Nozoe (b. 1902) was educated in Japan and taught at universities there and in Taiwan. His extensive research in the structure and synthesis of organic chemicals won him awards in Japan and a worldwide reputation. He tells the story of his boyhood, education, and career with much gracious praise for colleagues. The passages describing his research use chemical notation extensively and require advanced training in the discipline. Photos, drawings, and diagrams; end notes; indexes. 267 pp.

Patterson, Elizabeth C. *John Dalton and the Atomic Theory*. Garden City, N.Y.: Doubleday, 1970.
Biography. Patterson provides a succinct account of John Dalton's life and work. The great chemist rose from a poor childhood in England's Lake District to elucidate chemistry by classifying chemical properties based on the weight of atoms. But, as the author points out, Dalton also advanced the understanding of gases and heat, meteorology, and physiology. By the time of his death in 1844, his achievements earned wide fame, membership in prestigious societies, including the Royal Society, honorary degrees, and the patronage of royalty. Like Michael Faraday, Dalton came from a devout dissenting family (Quakers in Dalton's case) to become a leading scientist and lecturer. Portraits and diagrams; bibliography; end notes; index. 348 pp.

Pflaum, Rosalynd. *Grand Obsession: Madame Curie and Her World*. New York: Doubleday, 1989.
Biography. Marie Curie's story is an amazing one, and Pflaum tells it adroitly for general readers. Curie was born in Poland, became the first woman to earn a doctorate in physics in

France, collaborated with her husband—Pierre, also a genius at physics—in discovering polonium and radium, suggested the revolutionary notion that radioactivity (she coined the word) came from within atoms, helped develop radiation therapy for cancer, and won two Nobel Prizes, among many other awards. All the while, she and Pierre slowly poisoned themselves with radiation. They raised two daughters, one of whom, Irène, also won a Nobel Prize. Pflaum argues that incredibly simple tastes, physical toughness, and unswerving stubbornness and concentration made Curie a successful scientist in a day when women were expected only to marry and bear children. Photos; bibliography; index. 496 pp.

Pilkington, Roger. *Robert Boyle, Father of Chemistry*. London: John Murray, 1959.
Biography. Born to Irish aristocracy, Robert Boyle (1626-1691) was a staunch advocate of the atomic theory and derived the description of gases known as Boyle's law, but he contributed to many scientific subjects. His numerous publications through the Royal Society marked him as a leading anti-Aristotelian philosopher of his day, one who helped develop the scientific method. Pilkington's narrative gives about equal attention to Boyle's education, personal life, and intellectual pursuits. Diagrams; short bibliography; index. 179 pp.

Posin, Daniel Q. *Mendeleyev: The Story of a Great Scientist*. New York: McGraw-Hill, 1948.
Biography. Dmitry Ivanovitch Mendeleyev (1834-1907) devised the table of chemical elements. A great achievement of classification and synthesis, it allowed him to predict the existence of unknown elements, to the astonishment of nineteenth-century chemists. But he within Russia he was as well known for technological innovations. He helped develop mining and industry. Progressive politically, he championed the rights of the common person during the reign of the last Czar. Posin tells of Mendeleyev's youth in Siberia, his education and moves to St. Petersburg and Moscow, and his rise as Russia's most influential scientist. For general readers. Photos; bibliography; index. 345 pp.

Prelog, Vladimir. *My 132 Semesters of Chemistry Studies: Studium chymiae nec nisi cum morte finitur*. Washington, D.C.: American Chemical Society, 1991.
Scientific autobiography. The book's subtitle translates, "And the study of chemistry will not end except with death"—a testament to the obsession Vladimir Prelog (b. 1906) has for his profession. Born in Yugoslavia and educated in Prague, Prelog taught at universities in Prague, Zagreb, Zurich, and the United States. He helped create the Cahn-Ingold-Prelog (CIP) nomenclature system now used in chemistry and won the 1975 Nobel Prize for Chemistry because of his work in stereochemistry. Most of this brief book describes his research and university career for an audience of professional chemists, but the opening pages relate his boyhood and budding interest in science. Photos and diagrams; end notes; index. 120 pp.

Priestley, Joseph. *The Memoirs of Dr. Joseph Priestley*. Edited by John T. Boyer. Washington, D.C.: Barcroft Press, 1964.
Autobiography. This abbreviated edition of Joseph Priestley's autobiography (first published in 1806) is charming reading, although a sanitized account of his tempestuous, independent-minded life. He devotes only one short chapter to his experiments with air and electricity. The final third of the book, written by Priestley's son, tells of his religious persecution in England and his experiences in America. Short bibliography; index. 173 pp.

Roberts, John D. *The Right Place at the Right Time*. Washington, D.C.: American Chemical Society, 1990.
Scientific autobiography. John D. Roberts (b. 1918) describes his life as a researcher and teacher at the Massachusetts Institute of Technology and the California Institute of Technology. He specialized in the structure, synthesis, and use of organic chemistry, writing a textbook on the subject, but also worked on nuclear magnetic resonance spectroscopy. He briefly describes his youth and education, but most of the book involves his research at a level understandable only to professional chemists. Photos and diagrams; end notes; index. 299 pp.

Sayre, Anne. *Rosalind Franklin and DNA*. New York: W. W. Norton, 1975.
Biography. An X-ray crystallographer, Rosalind Franklin participated in the research that identified the structure of deoxyribonucleic acid (DNA), for which James D. Watson and Francis Crick won the Nobel Prize for Chemistry. Watson's popular book *The Double Helix* (1968) tells the story of the research. That book, argues Sayre, does an injustice to Franklin, who died before it was published. According to Sayre, Watson's version of Franklin is more fiction than fact, and she wrote this biography to explain what Franklin, whom she knew, was really like. Sayre devotes most of the narrative to careful explanations of Franklin's research and the extent of her contributions. Skillfully written, the book includes enough basic summary of molecular biology and biochemistry to give general readers an appreciation of DNA research and of a remarkable scientist. End notes. 221 pp.

Schofield, Robert E., ed. *A Scientific Autobiography of Joseph Priestley (1733-1804)*. Cambridge, Mass.: MIT Press, 1966.
Autobiography-biography. Schofield arranges the letters by Joseph Priestley to other scientists in order to reveal the development of his scientific ideas. Schofield provides an introduction and occasional commentary, but the writing is mostly Priestley's. Readers may find the text confusing unless they know something of life and science in the eighteenth century. Best read in conjunction with *The Memoirs of Dr. Joseph Priestley* (see above). Portrait of Priestley; long bibliography; index. 415 pp.

Serafini, Anthony. *Linus Pauling: A Man and His Science*. New York: Paragon House, 1989.
Biography. As Isaac Asimov comments in the foreword to this book and as Serafini demonstrates in the main text, one quality that stands out in Linus Pauling is force of character. He was a relentless man both in science and in his social views. A Nobelist in chemistry and peace, he led a varied, highly controversial life. Serafini takes up both facets in detail, explaining the science non-technically—chemical bonding, theory of anesthesia, pathology of sickle-cell anemia, structure of

DNA—and tracing the growth and assertion of his political ideas. Serafini also describes Pauling's private life. Photos; bibliography; end notes; index. 310 pp.

Staudinger, Hermann. *From Organic Chemistry to Macromolecules.* New York: Wiley-Interscience, 1970.
Scientific autobiography. Hermann Staudinger (1881-1965) proposed that polymers, organic substances of high molecular weight, could grow in chains of enormous size. The insight was crucial to development of the plastics industry, and in 1953 he was awarded the Nobel Prize for Chemistry because of it. He also trained many German professional chemists at universities in Zurich and Freiburg. This book touches only briefly on his rearing and private life before describing his research in technical detail accessible only to readers with college-level training in chemistry. Photos; end notes. 303 pp.

Susac, Andrew. *The Clock, the Balance, and the Guillotine.* Garden City, N.Y.: Doubleday, 1970.
Biography. Susac tells the story of Antoine Lavoisier's life through chapters devoted to the leading influences on him—members of his family, scientists and teachers, and his executioners during the French Revolution. The explanations of science are sketchy as the text focuses on political and social currents during Lavoisier's times. The presentation occasionally employs the techniques of fiction—such as dialogue—to dramatize key events. Index. 206 pp.

Tarbell, D. Stanley, and Ann Tracy Tarbell. *Roger Adams: Scientist and Statesman.* Washington, D.C.: American Chemical Society, 1981.
Biography. Roger Adams (1889-1971), the leading organic chemist of his generation according to the authors, was industrious both as a teacher and researcher. He trained more than 200 doctoral and post-doctoral students and conducted research in academia and industry. He also served the U. S. government efforts in both world wars. The authors seek to isolate those qualities in Adams that made him an outstanding teacher and scientist. Chemical notation is part of

the explanations of Adams' research. Photos; end notes. Index. 240 pp.

Thackray, Arnold. *John Dalton: Critical Assessments of His Life and Science*. Cambridge, Mass.: Harvard University Press, 1972.
Biography-history. John Dalton (1766-1844) advanced the chemical atomic theory, described color blindness, and improved the understanding of gases. In separate essays the author discusses Dalton's role in the Industrial Revolution, the development of atomic theory, his natural philosophy, and the development of modern chemistry. Two more chapters contain documents by Dalton concerning atomic theory and his scientific correspondence. Photos of documents and drawings; bibliographic essay; indexes. 190 pp.

Thorpe, Thomas Edward. *Joseph Priestley*. London: Dent, 1906.
Biography. Thorpe sets himself the task of showing how Joseph Priestley, the "honest heretic," was a hero and the epitome of intellectual energy in the eighteenth century. Most of the text concerns Priestley's activities in religion and politics, as well as his upbringing and family life. Only the last chapter takes up his scientific work, covering the study of gases—and the discovery of oxygen in particular—by quoting extensively from Priestley's publications. Drawings; index. 228 pp.

Travers, Morris W. *A Life of Sir William Ramsay*. London: Edward Arnold, 1956.
Scientific biography. Travers devotes most of this book to explaining the scientific discoveries of William Ramsay (1852-1916) in technical detail and placing them in the context of Ramsay's university career at Glasgow, Bristol, and University College, London. Principal among his achievements were his discoveries of the noble gases helium, argon, neon, krypton, and xenon, for which he won the 1904 Nobel Prize for Chemistry. He also studied radon and radium. Such work, along with his teaching, made him one of the most celebrated British chemists of his day. Readers will need at least a basic knowledge of chemistry to appreciate the text. Photos and diagrams; index. 308 pp.

Treneer, Anne. *The Mercurial Chemist: A Life of Sir Humphry Davy.* London: Methuen, 1963.

Biography. Although Treneer discusses the scientific research of Humphry Davy, devoting a chapter to his safety lamp, the mainline of the book concerns Davy's relation to poets and his personal life and travels. Davy knew William Wordsworth (his parody of the poet's style is included) and Coleridge and wrote poetry himself. He found delight in nature and had an extraordinary zest for life, which Treneer reveals admirably. The book supplements biographies that attend to Davy's science. Photos and drawings; bibliography; index. 264 pp.

Vare, Ethlie Ann. *Adventurous Spirit: A Story About Ellen Swallow Richards.* Minneapolis, Minn.: Carolrhoda Books, 1992.

Biography for readers ten years and older. Ellen Swallow Richards (1842-1911), according to Vare, was the first professional woman chemist in the United States, and the book naturally presents her as a courageous pioneer in a field traditionally dominated by men. During her career, Richards founded the study of home economics and studied water pollution. Emphasizing the dangers of pollution to health, she was in the vanguard of the environmental movement. Drawings; bibliography. 64 pp.

Williams, Trevor I. *Robert Robinson, Chemist Extraordinary.* Oxford, England: Clarendon Press, 1990.

Biography. Robert Robinson (1886-1975) had a long career as a professor of chemistry and served terms as president of the British Association for the Advancement of Science and the Royal Society. In these capacities he strongly influenced English science education. He also contributed to the electronic theory of chemical reactions. Williams, who knew Robinson and benefited from his influence, writes admiringly of him, concentrating on his academic career and administrative achievements. Photos; end notes; indexes. 201 pp.

Willstätter, Richard. *From My Life.* New York: W. A. Benjamin, 1965.

Autobiography. Richard Willstätter (1872-1942) earned the 1915 Nobel Prize for Chemistry for proving that photosynthesis involves two types of chlorophyll. Born, raised, and educated in Germany, he taught in Zurich and Munich and conducted research at the Kaiser Wilhelm Institute in Berlin-Dahlem. In 1924, only 52 years old, he resigned his professorship at the University of Munich to protest the treatment of Jews at the university; 15 years later he had to flee the Nazis by emigrating to Switzerland. A close friend of Fritz Haber, another Jewish refugee, as well as many other eminent chemists, Willstätter recounts the tenor of German social and scientific life in the first quarter of the twentieth century and also reflects upon his private life. For all readers interested in the history of science, but a basic knowledge of chemistry is helpful. Photos; end notes; indexes. 461 pp.

Chapter 4

Earth Sciences

COLLECTIONS

Colbert, Edwin. *The Great Dinosaur Hunters and Their Discoveries.* New York: Dutton, 1968; Dover, 1984.
Biography-history. Colbert writes with gentlemanly charm and humor about the amateurs and scientists worldwide who unearthed dinosaur skeletons and pieced together from them a view of climate and life during the Mesozoic era. The survey, full of biographical detail and anecdotes, begins with William Buckland in the early nineteenth century and ends in the 1960's with developments that occurred just before Colbert wrote the first edition. Many photos and drawings; bibliography; index. 283 pp.

Fenton, Carroll Lane, and Mildred Adams Fenton. *The Story of the Great Geologists.* Garden City, N.Y.: Doubleday, Doran, and Company, 1945; *Giants of Geology*, 1952.
Biography-history. The Fentons recount the rise of modern geology through the lives of naturalists, explorers, and scientists who described the terrains, rocks, and fossils they found. The authors begin with a brief look at geological theories from classical Greece and Rome and the Middle Ages. Then they take up discoveries and discoverers in more detail, from the Dane Nils Steensen (Nicolaus Steno, 1638-1686) to the American glaciologist Thomas Chrowder Chamberlain (1843-1928), with dozens between. For general readers. Photos and drawings; bibliography; index. 301 pp.

BIOGRAPHIES AND AUTOBIOGRAPHIES

Bailes, Kendall E. *Science and Russian Culture in an Age of Revolutions: V. I. Vernadsky and His Scientific School, 1863-1945.* Bloomington: Indiana University Press, 1990.
Biography. This book's ostentatious title accurately reflects the author's central interest in writing it—political history. Vladimir Ivanovich Vernadsky was a geochemist of great originality and influential in Czarist and Soviet academic and industrial geology. Yet he had little sympathy with either political system, according to the author, and under the Soviets was considered to be brilliant but political unreliable. Accordingly, some of his philosophical works about science suffered suppression. Although Bailes summarizes Vernadsky's scientific ideas well, the focus is on his uneasy relations with his leaders and his unsuccessful attempt to keep Russian science free of political controls. Photos; bibliography; end notes; index. 238 pp.

Bailey, Edward Battersby. *Charles Lyell.* Garden City, N.Y.: Doubleday, 1963.
Biography. A compact, stiffly written life of the father of modern geology. Charles Lyell (1797-1875) profoundly shaped nineteenth-century ideas about Earth's history with his *Principles of Geology* (1833); among those Lyell influenced was Charles Darwin. Bailey provides a chapter early in the narrative to explain the geological theories prevailing when Lyell was young. With this background, the achievement of Lyell's catastrophism theory and its relation to modern theory become clear. Photos and drawings; index. 214 pp.

Bailey, Edward Battersby. *James Hutton—The Founder of Modern Geology.* London: Elsevier Publishing, 1967.
Biography. A geologist and historian of geology, Bailey writes of the progenitor of his discipline, James Hutton (1726-1797). A Scottish physician and farmer, Hutton studied soils, erosion, and topography. He suggested that coal was the remnants of ancient plants, that forces at the earth's center affected the surface, and that geological features formed over millions of years. Such observations branded him as an atheist

to many of his contemporaries, but he influenced the ideas of many later earth scientists. Most of this volume concerns summaries and explications of Hutton's publications, notoriously convoluted in style. 161 pp.

Barkhouse, Joyce C. *George Dawson, the Little Giant*. Toronto, Canada: Clarke, Irwin and Company, 1974.
Biography for readers 14 years and older. George Dawson (1849-1901) is known as the father of Canadian anthropology for his early study of Northwest Native Americans, but as a member of the North American Boundary Commission, he also studied much of Canadian topography, fossils, and geology, particularly in British Columbia. Barkhouse tells his story dramatically, relying on exciting scenes and dialogue to capture young readers' interest. Photos; bibliography; index. 138 pp.

Bascom, Willard. *The Crest of the Wave.* New York: Harper and Row, 1988.
Autobiography. Casually written and loosely structured, this book nevertheless tells the story of a remarkable life. Trained as a mining engineer, Willard Bascom switched to oceanography in 1945. It was still an infant science, and he helped it grow. He worked on both civilian and military projects, including wave dynamics, detection equipment, deep sea drilling, salvage, and the development of scuba apparatus. His work took him all over the world, including the Bikini atoll, where he studied the waves generated by the first hydrogen bomb detonation. An intriguing resource for information on pioneer oceanographers and early research methods. Photos. 318 pp.

Berkeley, Edmund, and Dorothy Smith Berkeley. *George William Featherstonhaugh*. Tuscaloosa: University of Alabama Press, 1988.
Biography. George William Featherstonhaugh (1780-1866) was born in England and largely self-educated yet became an unusually daring and learned man. He was the first U.S. government geologist, and according to the authors, Featherstonhaugh's survey of the Maine-New Brunswick bor-

der helped avert a war with England. He also studied Native American languages, launched railroads, translated Dante's *Inferno*, and rescued King Louis Philippe of France from a revolution. A busy, colorful life, which the Berkeleys relate vividly in this carefully researched book. Photos, portraits, and drawings; bibliography; end notes; index. 357 pp.

Champlin, Peggy. *Raphael Pumpelly: Gentleman Geologist of the Gilded Age*. Tuscaloosa: University of Alabama Press, 1994.
Biography. Champlin relates the career of Raphael Pumpelly (1837-1923) in order to describe the developing discipline of geology in nineteenth-century America. A colorful figure, Pumpelly was a mining geologist and explorer, consultant to the governments of China and Japan, and leading figure in the U.S. Geological Survey. He contributed to the fields of petrology, geomorphology, glaciology, and structural geology but like most of his contemporaries was as interested in resource development, principally of minerals, as pure science. Champlin emphasizes Pumpelly's scientific interests but also discusses his travels and archeological expeditions in Asia. Photos; bibliography; end notes; index. 273 pp.

Clark, Robert D. *The Odyssey of Thomas Condon*. Portland: Oregon Historical Society Press, 1989.
Biography. Clark gives the reader a satisfying American success story. Born to a poor family in Ireland, Thomas Condon (1822-1907) immigrated to the United States, educated himself, became a missionary to Oregon, discovered fossils of horse precursors, wrote about evolution, and became a professor of geology at the state university. His career did have its setbacks, however. According to Clark, Condon had to battle East Coast professors to receive the credit he deserved for his scientific discoveries. A richly detailed book. Photos; end notes; index. 567 pp.

Clarke, John M. *James Hall of Albany, Geologist and Palaeontologist, 1811-1898*. New York: Arno Press, 1978.
Biography. In this reprint of the 1923 edition, Clarke recounts the 67-year career of James Hall, who wrote massive tomes on the geology and fossils of New York and the northeastern

states. A friend of Charles Lyell and Louis Agassiz, Hall was an influential figure in nineteenth-century American academic geology, especially the interpretation of rock strata, and was honored internationally for his research. Clarke quotes Hall's correspondence extensively. Photos; index. 565 pp.

Colbert, Edwin H. *A Fossil-Hunter' s Notebook*. New York: E. P. Dutton, 1980.
Autobiography. A widely traveled paleontologist and popular writer, Edwin H. Colbert (b. 1905) helped excavate fossils in the United States, Africa, South America, the Middle East, India, Australia, and Antarctica. He describes his boyhood in Missouri, education, and friendship with many of the great paleontologists of the twentieth century and his expeditions as a staff member of the American Museum of Natural History. His adventures make exciting reading as he describes the growth of paleontology from the 1920's through the 1960's. Photos and drawings; index. 242 pp.

Cole, Douglas, and Bradley Lockner. *The Journals of George M. Dawson: British Columbia, 1875-1878*. 2 vols. Vancouver: University of British Columbia Press, 1989.
Memoirs. The editors provide a biographical sketch of George M. Dawson. The rest of the two volumes contain journal entries from four seasons of Dawson's surveying expeditions in British Columbia, during which he also investigated the area's geology. Photos; biographical directory; bibliography; index.

Drake, Ellen Tan. *Restless Genius: Robert Hooke and His Earthly Thoughts*. New York: Oxford University Press, 1996.
Biography. Although Robert Hooke (1635-1703) is best known for his physical discoveries and inventions and his role as official experimenter for the fledgling Royal Society, he also wrote perceptively about geology, and earthquakes in particular. Drake's book contains two parts. The first part begins with a substantial general biography and then explains Hooke's geological observations and theories thoroughly. The second part is an annotated edition of Hooke's *Discourse of Earthquakes and Subterraneous Eruptions*. Although accessible to

determined general readers, science historians and earth scientists are best prepared to appreciate Drake's study of Hooke. Photos, drawings, and diagrams; bibliography; end notes; index. 386 pp.

Fairbanks, Helen, and Charles Berkey. *Life and Letters of R. A. F. Penrose, Jr.* New York: Geological Society of America, 1952.
Biography. A memorial volume about a large benefactor to the Geological Society of America, Richard Alexander Fullerton Penrose (1827-1908). The authors admit the book would not have been written had not the millionaire Penrose given a lot of money to the society, where they were staff members. In any case, Penrose became rich as a geologist working in rich mining districts in the West, such as Cripple Creek, Colorado. He helped establish the methods for surveying and exploiting ore deposits. The book is incredibly overdetailed, but out of the mass of information patient readers can discern the mentality and industry of nineteenth-century mine developers. Letters constitute much of the text. Photos. 765 pp.

Foster, Mike. *Strange Genius: The Life of Ferdinand Vandeveer Hayden.* Niwot, Colo.: Roberts Rinehard Publishers, 1994.
Biography. A rival of Clarence King and John Wesley Powell for the first directorship of the U.S. Geological Survey, Ferdinand Hayden (1829-1887) led extensive survey expeditions in the West, including Yellowstone. He became well known as a collector of rocks, fossils, and wildlife specimens, helping such leading nineteenth-century naturalists as Asa Gray and Joseph Hooker. But according to Foster, Hayden was his own worst enemy. An irascible man, he performed poorly before politicians and was easy prey for the sharper Powell in winning government office. His story reveals both the nature of earth sciences in post-Civil War America and the political background for the development of the West. Photos; bibliography; end notes; index. 443 pp.

Friedman, Robert Marc. *Appropriating the Weather: Vilhelm Bjerknes and the Construction of Modern Meteorology.* Ithaca, N.Y.: Cornell University Press, 1989.

Biography-history. Vilhelm Bjerknes (1862-1951) founded and led the "Bergen school" of meteorology, named after its base in western Norway, and Bjerknes himself has become known as the father of modern meteorology. His innovation rested on applying principles of physics to weather forecasting, including investigations of high-altitude air. His school successfully modeled cyclone evolution and discovered the effects of polar air masses—achievements that coincided with the rise of aeronautics and modern agriculture, both of which needed accurate weather forecasting. Friedman's book is a scholarly assessment of Bjerknes and his school; the two principal purposes are to show why Bjerknes abandoned theoretical physics for meteorology and how he successfully spread his weather forecasting methods worldwide. A fascinating, thorough study of an underexamined area of modern science history, accessible to general readers with a basic knowledge of science. Photos, drawings, and diagrams; footnotes; index. 251 pp.

Geikie, Archibald. *A Long Life's Work*. London: Macmillan, 1924.
Autobiography. Sir Archibald Geikie (1835-1924) was both a geologist and a historian of geology. As a geologist, he was a professor at the University of Edinburgh, where he studied the structure of the Scottish highlands, and Director-General of the Geological Survey of England, Scotland, and Ireland. From 1908 to 1913 he served as president of the Royal Society. His long career saw the beginnings of modern geology, which he describes in a stiffly formal but manageable style for general readers. Photos; index. 426 pp.

Gerstner, Patsy. *Henry Darwin Rogers, 1808-1866: American Geologist*. Tuscaloosa: University of Alabama Press, 1994.
Biography. The son of an Irish revolutionary, Henry Darwin Rogers was a determined, strong-minded man who devoted himself to American geology, then an embryo science. He helped develop it in extensive surveys in Pennsylvania and New Jersey. His work there and his theory of mountain development, formulated while studying the Appalachians, brought him early fame, but his intellectual rigidity and harshness with colleagues led to his disfavor nearly as

quickly. Gerstner's biography of this influential pioneering geologist affords insights into the state of the science and the temperaments of scientists before the Civil War. Drawings and diagrams; bibliography; end notes; index. 311 pp.

Gould, Charles N. *Covered Wagon Geologist*. Norman: University of Oklahoma Press, 1959.
Autobiography. Charles N. Gould (1868-1949) had a 60-year career investigating the geology and paleontology of the Southwest, and his story is distinctively American. Born in a log cabin, he earned his way through college, became a professor at the University of Oklahoma, conducted the state geological survey, and worked as a geologist for the National Park Service. He helped discover the Oklahoma oil fields, wrote hundreds of papers about geology, and won honorary degrees and memberships in learned societies both in the United States and elsewhere. Photos and maps; index. 282 pp.

Hendrickson, Walter Brookfield. *David Dale Owen: Pioneer Geologist of the Middle West*. Indianapolis: Indiana Historical Bureau, 1943.
Biography. David Dale Owen (1807-1860) was a surveyor and chemist in the mid-nineteenth century who studied the rock structure, fossils, and mineral deposits of the Upper Mississippi Valley. Well known by colleagues in his day, Owen was an obscure figure by the time Hendrickson's book was published, and the author's intention was to rescue Owens from that obscurity. For general readers interested in regional history, Hendrickson offers a plodding, thorough review of Owen's career. Photos and drawings; bibliography; footnotes; index. 180 pp.

Higham, Norman. *A Very Scientific Gentleman*. Oxford, England: Pergamon Press, 1963.
Biography. Higham's biography of Henry Clifton Sorby (1826-1908) depicts a typical scientist of Victorian England. Independently wealthy, Sorby worked on problems that interested him, and as soon as he had satisfied himself he had an answer, he drifted on to other problems. His understanding was often profound. He helped found metallography, clar-

ifying the microscopic structure of steel and granite, and petrology. Retiring by nature, he published more than 200 papers, many in a literary society in his native Sheffield. In a style suited to general readers interested in science history, Higham surveys Sorby's many intellectual achievements and considers the origin and elements of his temperament. Photos; bibliography; end notes; index. 160 pp.

Jaggar, Thomas A. *My Experience with Volcanoes.* Honolulu: Hawaiian Volcano Research Association, 1956.
Autobiography. Thomas A. Jaggar (1871-1953) was a pioneer vulcanologist who began his research in Hawaii in 1912 after earning a degree in geology at Harvard University and applying his knowledge on an expedition to Mount Vesuvius and on research of California earthquake faults and the Aleutian islands. Dangerous work all of it, but Jaggar describes his career with matter-of-factness. He goes into considerable detail about the fine points of research, the behavior of volcanoes, and his colleagues. Photos; index. 198 pp.

Le Conte, Joseph. *The Autobiography of Joseph Le Conte.* Ed. by William Dallam Armes. New York: D. Appleton and Company, 1903.
Autobiography. Initially trained as a physician, Joseph Le Conte (1823-1901) returned to school at Harvard to study geology under Louis Agassiz. Thereafter he taught at universities in Georgia, South Carolina, and California, where he directed the Department of Geology and Natural History. In this book, completed shortly before his death, he recalls his family life, studies, service as a chemist in the Confederate Army, and geological research in the Sierra Nevada and other areas in the West. In 1891 he became president of the American Association for the Advancement of Science, the high point of his long career. Photos. 337 pp.

Lessem, Don. *Jack Horner: Living with Dinosaurs.* New York: W.H. Freeman, 1994.
Biography for readers ten years and older. Lessem offers a lively, inspiring account of the career of John R. Horner, an unusual success story. Dyslexic, Horner struggled in school

and never finished college, but his passion for collecting dinosaur fossils, which began at age seven, led him to jobs as museum curator and teacher, honorary degrees, and fame among paleontologists, especially for his discovery that some dinosaurs were warm blooded. His extensive discoveries about how dinosaurs lived inspired a central character in the movie *Jurassic Park*, for which he was a science consultant. Many drawings; short bibliography; index. 48 pp.

McGrath, Sylvia Wallace. *Charles Kenneth Leith, Scientific Advisor.* Madison: University of Wisconsin Press, 1971.
Biography. The author presents Charles Kenneth Leith (1875-1956) as the prototype of the twentieth-century scientist-consultant. A University of Wisconsin professor, Leith earned his reputation first as a structural geologist specializing in the Precambrian era. He also consulted for industry and served on government boards during the Roosevelt administration, playing a central role in procuring minerals during World War II. Photos; bibliographic essay; index. 255 pp.

Munson, Richard. *Cousteau: The Captain and His World.* New York: William Morrow, 1989.
Biography. Jacques Cousteau (1910-1997) is one of the most famous names on earth, thanks in large part to his genius for making films about the underwater world. To Munson, film-making is in fact Cousteau's greatest achievement, seconded by his innovations in underwater technology, especially the breathing device that made extended diving possible. Cousteau's films and scuba opened the oceans to curious non-scientists and awakened the public to dangers faced by marine animals and the environment from industrialized society. Cousteau's scientific achievements are more controversial. In his mildly critical biography, Munson shows Cousteau to be more a self-promoter and facilitator than a scientist or environmentalist. The book is enjoyable reading if only because Cousteau successfully led the life of a maverick adventurer. Photos; end notes; index. 316 pp.

Newbigin, Marion L., and J. S. Flett. *James Geikie.* Edinburgh, Scotland: Oliver and Boyd, 1917.

Biography. The younger brother of Sir Archibald Geikie, James Geikie (1839-1915) worked for the Geological Survey of Scotland and followed his brother to a professorship at the University of Edinburgh. He studied glaciers and proposed that a great ice age had sculpted many of the earth's topographical features. Half of this book treats Geikie's personal life and career, and half explains his theories about glaciers. Plainly written for all readers interested in nineteenth-century science. Photos; bibliography; index. 227 pp.

Pettijohn, Francis John. *Memoirs of an Unrepentant Field Geologist*. Chicago, Ill.: University of Chicago Press, 1984.
Autobiography-history. F. J. Pettijohn (b. 1904) tells about his life and career as a way of explaining the fertile expansion of geology following World War II. He contributed to the development of sedimentology and geochemistry and was on the periphery of those who promulgated plate tectonics. He explains his part in these subdisciplines and those of others, including many anecdotes about colleagues. Not an orderly history, the book is nevertheless an insight into the growth of a science from the point of view of a participant. Photos and diagrams; bibliography; index. 260 pp.

Pumpelly, Raphael. *My Reminiscences*. 2 vols. New York: Henry Holt, 1918.
Autobiography. A remarkable life told with humorous verve and a descriptive style similar to that of Mark Twain. Pumpelly studied the Ice Age in Europe, worked on mines in the Southwest during the Apache war, served as a mining and geology advisor to the Japanese and Chinese governments, toured throughout Asia, worked for the U.S. Geological Survey and Tenth Census, and organized the Northern Transcontinental Survey. Readers interested in nineteenth-century life, travel, and geology will prize this book. Photos, drawings, and diagrams; index in Vol. 2.

Scholander, Per Fredrik. *Enjoying a Life in Science*. Fairbanks: University of Alaska Press, 1990.
Autobiography. Per Scholander (1905-1980) was a versatile physiologist, working at various times on the physiology of

diving, glaciers, osmosis, and the response of animals to low temperatures. Born in Sweden and raised and educated in Norway, Scholander traveled widely on the oceans during his career, and he describes his many adventures in such places as the Arctic, Greenland, Brazil, Alaska, and the Amazon. Unfortunately, he was not a very skilled writer, at least in this volume put together from his manuscripts by his colleague Robert Elsner. The book is a series of loosely connected anecdotes, engaging in themselves but containing little beyond personal impression, and he does not explain his scientific investigations in detail. Photos; bibliography; index. 226 pp.

Schwarzbach, Martin. *Alfred Wegener, the Father of Continental Drift*. Madison, Wis.: Science Tech, 1986.
Biography. Alfred Wegener (1880-1930) drew scorn from contemporary geologists by proposing in 1912 that the continents constantly slide over the Earth's surface; his idea gained wide acceptance as part of plate tectonics theory in the late twentieth century. As Schwarzbach shows, Wegener was far more than a theorist. As a professor of meteorology and geophysics at Graz, Austria, he was an early investigator of the jet stream and paleoclimatology. He led four expeditions to explore Greenland, dying on the last one. As a young man, he even set a ballooning record with his brother. The author opens the book with a short chapter on Alexander von Humboldt, Wegener's intellectual forebear, then discusses his life and career, Greenland explorations, geophysical ideas, and the rise of plate tectonics. Although the primary audience appears to be geologists, any reader interested in science can enjoy this book's account of a romantic figure. Photos, drawings, and diagrams; bibliography; end notes; index. 241 pp.

Shortland, Michael, ed. *Hugh Miller's Memoirs: From Stonemason to Geologist*. Edinburgh, Scotland: Edinburgh University Press, 1995.
Biography-memoirs. "History has not been kind to Hugh Miller" (1802-1856), Shortland writes. Yet this newspaperman, controversialist, poet, and geologist had a great impact on early Victorian England. A long introduction by Shortland

explains Miller's importance as he summarizes his life and theories about ancient life based upon fossils found in Scotland. The rest of the text comprises two letters in which Miller reflects upon his early career. Bibliography; end notes; index. 266 pp.

Stafford, Robert A. *Scientist of Empire: Sir Roderick Murchison, Scientific Exploration and Victorian Imperialism.* Cambridge, England: Cambridge University Press, 1989.
Biography. To Stafford, the career of Roderick Murchison (1792-1871) illustrates the relation between science and England's expansionist political policies during the nineteenth century. As director-general of the Geological Survey of Great Britain, Murchison worked to institutionalize natural sciences by offering the fruits of research and exploration to policy makers to help them identify lucrative areas worldwide for exploitation. Only Stafford's first chapter is biographical; the following six chapters analyze the effects Murchison had on the culture of various regions of the world. The biographical chapter examines Murchison's contribution to the understanding of geological strata and topographical features. Photos and drawings; bibliography; end notes. 293 pp.

Stegner, Wallace. *Beyond the Hundredth Parallel: John Wesley Powell and the Second Opening of the West.* Boston, Mass.: Houghton Mifflin, 1953.
Biography. John Wesley Powell (1834-1902) led the first exploration team through the Grand Canyon, was the guiding spirit of the U.S. Geological Survey and its second director, studied and classified Native American languages, and campaigned for reform of land-use laws in the West as the most powerful scientist in the federal government during the 1870's and 1880's. He was also, according to Stegner, the father of physical geology and a perceptive hydrologist and topographer whose findings about erosion are particularly valuable. Stegner explains Powell's geological innovations superficially in order to concentrate on the drama of Powell's harrowing trips down the Colorado River and his battles—harrowing politically—with Congress and jealous rival scientists. The book is an absorbing study of science and politics in late-nine-

teenth-century America. Photos and drawings; end notes; index. 438 pp.

Stephens, Lester D. *Joseph LeConte, Gentle Prophet of Evolution.* Baton Rouge: Louisiana State University Press, 1982.
Biography. Born and reared on a plantation in the antebellum South, Joseph Le Conte (1823-1901) had to flee the Reconstruction after the Civil War. He secured a professorship in zoology and geology at the University of California, a fledgling institution, and after long consideration accepted Charles Darwin's theory of evolution. Much of Le Conte's professional career was devoted to attempts to reconcile evolution and Christian tenets; his many books, lectures, and articles made him a controversial figure. He also helped describe the physiology of the eye and served as president of the American Association for the Advancement of Science. Stephens places Le Conte in the context of his times in order to clarify the intellectual's life in nineteenth-century America. Photos; bibliography; footnotes; index. 340 pp.

Terrell, John Upton. *The Man Who Rediscovered America.* New York: Weybright and Talley, 1969.
Biography. Terrell's biography of John Wesley Powell gives only a passing glance at his contributions to geology and ethnology. Terrell concentrates on Powell's exploration of the West, his tenure as director of the U.S. Geological Survey, and his losing battle with Congress to develop western lands in such a way that settlers would have a fair chance of succeeding on their homestead claims. Bibliography; index. 281 pp.

Vaughan, Francis E. *Andrew C. Lawson: Scientist, Teacher, Philosopher.* Glendale, Calif.: Arthur H. Clark Company, 1970.
Biography. A prominent geologist and professor at the University of California, Andrew C. Lawson studied the geology of California and Nevada extensively. He also worked on the Geological Survey of Canada as a young man. This volume is a memorial by his many students, recounting his career and summarizing the contents of his primary publications. The book concludes with a selection of Lawson's poems,

anecdotes illustrating his sharp wit, and testimonials from academics and industrial geologists. Photos; index. 474 pp.

Wertenbaker, William. *The Floor of the Sea: Maurice Ewing and the Search to Understand the Earth*. Boston, Mass.: Little, Brown, 1974.
Biography. Maurice William Ewing (1906-1974) explored the Earth's crust, especially the sea floor and the mid-ocean ridge. He became a proponent of plate tectonics when the theory was intensely controversial and uncovered sea fossils and living creatures deep in oceanic trenches. One of the pioneers of oceanography, he conducted soundings and samplings on many voyages, as Wertenbaker relates in this science-centered biography. The writing is not lively, but Wertenbaker explains the details of Ewing's research in detail. Photos and drawings; bibliography; index. 275 pp.

Westman, Paul. *Jacques Cousteau*. Minneapolis, Minn.: Dillon Press, 1980.
Biography for readers eight years and older. Westman writes simply about the career of Jacques-Ives Cousteau. The book shows Cousteau to be inquisitive, hardworking, and daring as he learns to dive, helps invent the scuba breathing device, and films undersea life worldwide, for the first time making the oceans a spectacle for everyone. Westman also touches upon Cousteau's efforts to protect marine life from pollution. Photos and drawings. 47 pp.

Wilkins, Thurman. *Clarence King: A Biography*. New York: Macmillan, 1958; revised and enlarged ed. with the help of Caroline Lawson Hinkley, Albuquerque: University of New Mexico Press, 1988.
Biography. The life of Clarence King (1842-1901) was unquestionably among the most vivid, brilliant, and tragic in nineteenth-century American science. To his friends Henry Adams and John Hay, King was the best and brightest mind of his generation, unsurpassed in wit, energy, and charm. King climbed many Sierra Nevada peaks, conducted geological surveys of exceptional accuracy in California and along the fortieth parallel, was a cattle baron in Wyoming, and helped

conceive and was the first director of the U.S. Geological Survey. He contributed original data on and theoretical explanations of glaciers, mountain formation, and ore location, becoming the best mining geologist in the country. But he abandoned pure science in the 1880's in a fruitless search for wealth and frittered away his talents. Although he charmed the great intellectuals of Britain and America—John Ruskin, William Dean Howells, and Henry James among them—he died without wealth and without fulfilling the great potential everyone saw in him. He also led a double life, married secretly to an African American. Wilkins' biography relates all of this complex, brilliant man's scientific interests, great zest for life, and intense ambition with great clarity, stylish prose, and perplexity, for to Wilkins, as to Henry Adams, King poses a particularly American conundrum of talent clashing with ambition; unfortunately, the biography also shows a delicacy of treatment that may make readers wonder whether Wilkins has avoided some topics. Nevertheless, among biographies of scientists, this is a classic for all readers. Photos; bibliography; end notes. 524 pp.

Wilson, Leonard G. *Charles Lyell, the Years to 1841: The Revolution in Geology*. New Haven, Conn.: Yale University Press, 1972.
Biography. Wilson delivers a close study of the youth and early career of Charles Lyell. He describes Lyell's studies at Oxford University, the state of geology at the time (dominated by catastrophism), his travels in Europe and America, first publication of *The Principles of Geology* (1830-1832) and subsequent editions, his relations with Charles Darwin, and his theory of glaciers. The uniformitarian theory of Lyell revolutionized geology, as Wilson explains. An editor of Lyell's letters, he quotes from them frequently. A scholarly biography , the book is a useful resource on nineteenth-century geology for all readers interested in science history. Portraits, drawings, and diagrams; bibliography; footnotes; index. 553 pp.

Youngquist, Walter. *Over the Hill and Down the Creek*. Caldwell, Idaho: Caxton Printers, 1966.
Memoirs. Youngquist recounts his experiences as a field geologist and many anecdotes about colleagues working in far-

flung locales to locate the minerals and petroleum for American consumption. The tone is humorous and the discussions are non-technical, although a good deal of geology receives attention. A good book for students thinking about becoming geologists, although it was published when exploiting Earth's resources was held to be a much more heroic occupation than during the subsequent decades. Drawings. 322 pp.

Chapter 5

Life Sciences

COLLECTIONS

Adams, Alexander B. *Eternal Quest: The Story of the Great Naturalists*. New York: G. P. Putnam's Sons, 1969.
Biography-history. Although Adams discusses some geologists and astronomers, he concerns himself mainly with the natural philosophers and scientists who most changed humanity's view of its relation to nature. He begins with Aristotle and ends with Gregor Mendel. The greater part of the text covers eighteenth- and nineteenth-century naturalists, particularly Carl Linnaeus, Georges Buffon, Jean Lamarck, Georges Cuvier, John James Audubon, Alexander von Humboldt, Charles Lyell, Louis Agassiz, Charles Darwin, Alfred Wallace, and Thomas Huxley. Adams provides fairly sanitized accounts of their personal lives and describes in broad terms their major achievements, all in support of his theme that such careers demonstrate "the courage of man and the dignity of his spirit." Photos and drawings; appendix with biographical sketches; bibliography; index. 509 pp.

The Biographical Dictionary of Scientists: Biologists. New York: Peter Bedrick Books, 1984.
After an introduction to the history of biology, 190 alphabetized entries provide basic biographical information and summaries of the careers of biologists, naturalists, and medical researchers (such as Galen) for a general audience. The coverage extends from Hippocrates to scientists now living. Articles average about 500 words. Drawings; glossary; index. 182 pp.

Bonta, Marcia Myers. *American Women Afield*. College Station: Texas A&M University Press, 1995.
This is a valuable collection of short biographical sketches and excerpts from the writings of 25 women naturalists, some of whom are featured in no other biographical collection. The editor starts with Susan Fenimore Cooper (1813-1894, daughter of the novelist James Fenimore Cooper) and concludes with Rachel Carson (1907-1964). The prefatory sketch for each writer summarizes her publications and contribution to the life sciences. Among the subjects are ornithologists, botanists, entomologists, and ecologists. Photos; bibliography for each writer and general bibliography. 248 pp.

Coates, Ruth Allison. *Great American Naturalists*. Minneapolis, Minn.: Lerner Publications, 1974.
For readers 12 years and older. Coates begins with a short introduction explaining what a naturalist is. Then she sketches the lives and accomplishment of John Bartram and William Bartram (1699-1777, 1739-1823), Alexander Wilson (1766-1813), John James Audubon (1785-1851), Henry David Thoreau (1817-1862), John Muir (1838-1914), John Burroughs (1837-1921), Luther Burbank (1849-1926), Theodore Roosevelt (1858-1919), Ernest Thompson Seton (1860-1946), George Washington Carver (1860-1943), and Rachel Carson. Photos and drawings. 103 pp.

De Kruif, Paul. *Microbe Hunters*. New York: Harcourt, Brace, 1926; 1996.
Biography-history. A classic of popular science, this collection first appeared during the early development of microbiology. De Kruif writes vividly and dramatically about the thinkers who created the new science and turned it into a powerful medical tool, thinkers that to him are intellectual heroes: Antony Leeuwenhoek, Lazzaro Spallanzani, Louis Pasteur, Robert Koch, Emile Roux, Emil Behring, Elie Metchnikoff, Theobald Smith, David Bruce, Ronald Ross, Battista Grassi, Walter Reed, and Paul Ehrlich. Although de Kruif's portraits lack depth, they exude great enthusiasm for facts and theories and portray the interplay among scientists. Drawings; index. 357 pp.

Dewsbury, Donald A. *Leaders in the Study of Animal Behavior.* Lewisburg, Pa.: Bucknell University Press, 1985.
To fill a gap he found in biographical information about animal behavior researchers, Dewsbury collected 19 autobiographical statements from leaders in the field. All are well known to colleagues, and a few, such as Konrad Lorenz and Edward O. Wilson, are famous to the general public for their popular books. International in scope, the book offers readers firsthand views of animal research and explanations of its purpose and techniques. Photos; short bibliography concluding each chapter. 512 pp.

Elman, Robert. *First in the Field.* New York: Mason/Charter, 1977.
In his introduction, Elman remarks upon the motivations and temperament of naturalists and the periodic resurgence of concern for nature. The second chapter surveys the forerunners of the eight men that he profiles in separate chapters. They are Mark Catesby (1682-1749), John and William Bartram, Alexander Wilson (1766-1813), John James Audubon, Louis Agassiz (1807-1873), John Wesley Powell (1834-1902), and John Burroughs (1837-1921). The mixture of botanists, zoologists, and geologists reflects Elman's definition of a naturalist as someone with an insatiable desire to understand this still-mysterious planet, a desire satisfied by observing nature in the field. Elman writes with grace for all readers. Photos and drawings; bibliography; index. 231 pp.

Faber, Doris, and Harold Faber. *Nature and the Environment.* New York: Charles Scribner's Sons, 1991.
Suited to readers 12 years and older. This volume belongs to the publisher's Great Lives series, which aims to encourage young people with the examples of success stories. Arranged in alphabetical order, the collection contains a mixture of scientists, philosophers, environmentalists, and government officials, including Louis Agassiz, Luther Burbank, Rachel Carson, George Washington Carver, Jacques-Yves Cousteau, Carl Linnaeus, John Muir, Gifford Pinchot, Theodore Roosevelt, Henry David Thoreau, and others. Photos and drawings; bibliography; index. 296 pp.

Humphrey, Harry Baker. *Makers of North American Botany.* New York: Ronald Press, 1961.
Humphrey summarizes the life and research of 121 American botanists. Arranged alphabetically, the entries average about 500 words in length. The criteria for inclusion were that the subject was dead at the time Humphrey compiled the collection and that each made a substantial or pioneering contribution to the identification, description, collection, or taxonomy of North American plants. The selections come from the seventeenth through the twentieth centuries, and Humphrey rescued many of them from obscurity for general readers. Brief bibliography for each entry. 265 pp.

Isely, Duane. *One Hundred and One Botanists.* Ames: Iowa State University, 1994.
Isely organizes his portraits of botanists chronologically, beginning with Aristotle (384-322 B.C.E) and finishing with Winona Hazel Welch (1896-1991). Much of his information came from secondary sources, and he selected those botanists who are known for major contributions. The brief entries (about 1,000 words each) emphasize the character and botanical work of the person. Despite an occasional flippancy of tone, the text contains much information for general readers. Photos and portraits; short bibliography for each entrant; indexes. 351 pp.

Kinkead, Eugene. *Spider, Egg, and Microcosm.* New York: Alfred A. Knopf, 1966.
Containing three essays originally published in *The New Yorker,* this book is delightful to read (including the sprightly, shrewd introduction by E. B. White). Based on interviews, Kinkead writes about three men in love with the subject of their research: arachnologist Alexander Petrunkevitch, embryologist Alexis Romanoff, and microbiologist Roman Vishniac. That all three are expatriate Russians brings the savor of politics to their stories, which, as White puts it, add "fuel to the fires of wonderment." 244 pp.

Miall, L. C. *The Early Naturalists.* London: Macmillan, 1912.
Miall opens with a brief essay about the study of natural his-

tory from classical times until the sixteenth century. Then he recounts the lives and work of naturalists from 1530 to 1789, more than 35 men in all. Nine sections organize the brands of naturalists: biologists, such as Leonhart Fuchs (1501-1566) and Guillaume Rondelet (1507-1566); the natural history of distant lands; early English naturalists, such as John Caius (1510-1573); John Ray (1627-1705) and his followers; anatomists of microorganisms, such as Robert Hooke (1635-1703) and Antony van Leeuwenhoek (1632-1723); pioneers of comparative anatomy, principally Claude Perrault (1613-1688); René-Antoine Ferchault de Réaumur (1683-1757) and his school; Carl Linnaeus (1707-1778), Bernard de Jussieu (1699-1777), and Antoine Laurent de Jussieu (1748-1836); and Georges Louis Leclerc, Comte de Buffon (1707-1788). Footnotes; index. 396 pp.

Oliver, F. W., ed. *Makers of British Botany*. Cambridge, England: Cambridge University Press, 1913.
This volume contains a general introduction by Oliver summarizing the work of botanists discussed later and commenting on the development of botany. Then academic botanists write 16 substantial portraits about their most famous forebears, including Nehemiah Grew, William Hooker, and John Lindley. One chapter sketches the professors of botany in Edinburgh from 1670 until 1887. The entries are written stodgily, but non-botanists interested in the subject can follow them. Photos and portraits; index. 332 pp.

Shteir, Ann B. *Cultivating Women, Cultivating Science: Flora's Daughters and Botany in England 1760-1860*. Baltimore, Md.: Johns Hopkins University Press, 1996.
Biography-history. The emphasis of this book is upon the participation of women in botanical studies and publications during the eighteenth and nineteenth centuries, but Shteir supplies much biographical information about women, writers especially, who are now little known. Shteir proposes that botany served as a significant part of the intellectual world and economy for women until in the nineteenth century botany became increasingly viewed as a "serious" scientific study for men to develop instead of an "amusement for

ladies," at which juncture women's role in botanical studies became problematical. Photos and drawings; bibliography; end notes; index. 301 pp.

BIOGRAPHIES AND AUTOBIOGRAPHIES

Adair, Gene. *George Washington Carver*. New York: Chelsea House Publishers, 1989.
Biography for readers 12 years and older. George Washington Carver (1864-1943) was a true American hero during his own lifetime: he came from a poor family of former slaves, left home to seek education, went to college despite the obstacles of racial prejudice, became a respected scientist, and received honors from nearly every segment of society. However, Adair shows that much about Carver derives from the media's myth-making after Carver became famous in the 1920's. Although he performed original research in agricultural science, his scientific achievements, especially those related to peanuts, were exaggerated. According to Adair, Carver's most important contribution was as an educator, mentor to both black and white students, and helper of poor farmers, and for these activities during his fifty-year career at the Tuskegee Institute in Alabama, the fame is amply merited. Photos; bibliography; index. 110 pp.

Adams, Alexander B. *John James Audubon*. New York: G. P. Putnam's Sons, 1966.
Biography. Adams opens this life of John James Audubon (1785-1851) with a description of the American wilderness, and its abundant bird life, before Europeans arrived to chop down the trees and harvest the fauna. Only a few far-sighted people, Adams writes, realized that a man was needed to record the original abundance before it was gone. That man was Audubon, according to Adams. He traces Audubon's illegitimate origins, youth, travels, and studies of birds as a lead-up to the publication of Audubon's great *Birds of America* from 1826 to 1836. A well-told, thorough account for general readers. Photos and drawings; bibliography; end notes; index. 510 pp.

Adams, Mark B., ed. *The Evolution of Theodosius Dobzhansky*. Princeton, N.J.: Princeton University Press, 1994.
Biography-history. Theodosius Grigorievich Dobzhansky (1900-1975) was a Russian-born entomologist and geneticist who, according to the editor, started population genetics in the United States. His professional and popular books, especially *Genetics and the Origin of Species* (1937), exerted great influence on American biological studies. His research and influence are the subjects of the 16 essays in this book, written by scholars primarily for scholars. Photos; bibliographies for each esssay; footnotes. 249 pp.

Ainley, Marianne Gosztonyi. *Restless Energy: A Biography of William Rowan, 1891-1957*. Montréal, Canada: Véhicule Press, 1993.
Biography. A colorful figure, William Rowan earned a secure place in the history of ornithology for his pioneering studies of bird migrations. He also made a name as a wildlife artist—a skill developed to help support his family because his professorship at the University of Alberta was not lucrative—but he preferred science. Trained in Europe, he found little administrative support for his specialty when he went to Canada to teach. For a general audience, Ainley explains the nature of his research and his concern about contemporary social issues. Photos and drawings; bibliography; end notes; index. 368 pp.

Allan, Mea. *The Hookers of Kew*. London: Michael Joseph, 1967.
Biography. Allan writes of William Jackson Hooker (1785-1865) and his son Joseph Dalton Hooker (1817-1911), two great botanists of nineteenth-century England. The father earned lasting fame for creating the lavish royal gardens at Kew. His son expanded the collection by exploring worldwide for new specimens—the Antarctic, the Himalayas, America, Australia, Palestine, and India—and helped persuade Charles Darwin to publish his theory of natural selection. Their long lives span the rise of botany as a formal scientific specialty, a development they fostered. The establishment of Kew by the first Hooker and the second's world travels in search of new plants for the gardens are Allan's main concern, but the tax-

onomy of plants also receives attention. Photos and drawings; bibliography; index. 273 pp.

Archer, Jules. *Science Explorer: Roy Chapman Andrews*. New York: Julian Messner, 1968.
Biography for readers 14 years and older. Roy Chapman Andrews (1884-1960) was a zoologist and the epitome of the intrepid explorer. He sailed on whalers in the north Pacific in order to bring back the first detailed studies and photographs of whales. He led five famous expeditions in China and Mongolia, which discovered fossils of dozens of new prehistoric species and the first dinosaur eggs. His world renown put him into the company of the social elite of the United States, and he became director of the Museum of Natural History. Archer tells Andrews' story with drama and verve, and an inspiring story it is for young readers, even if the era of such exploits on land is over. Bibliography; index. 191 pp.

Astro, Richard. *Edward F. Ricketts*. Boise, Idaho: Boise State University Press, 1976.
Biography. Edward F. Ricketts (1897-1948) became famous as the character Doc in John Steinbeck's novel *Cannery Row*. But as Astro points out, this notoriety obscures the considerable contribution that Ricketts made to marine biology. In particular, he published *Between Pacific Tides* (1939, with Jack Calvin), a pioneering guide to invertebrates of the central Pacific coast, and *Sea of Cortez* (1941, with Steinbeck). In this sketchy summary of Ricketts' career and eccentric intellect, Astro concentrates on Ricketts' influence on Steinbeck. Bibliography. 48 pp.

Ayres, Clarence. *Huxley*. New York: W. W. Norton, 1932.
Biography. Ayres emphasizes the drama in the life and long career of Thomas Henry Huxley (1825-1895). Certainly, there is plenty of drama as he recounts Huxley's reaction to the publication of Charles Darwin's *The Originof Species* (1859), zoological studies during a voyage aboard HMS *Rattlesnake*, educational reform, and vigorous public defense of Darwinism. The author weaves in important features of Huxley's private life, but his career always holds center stage. A readable account, the book clearly aims to humanize

Huxley while depicting him as an intellectual hero, especially to American readers. Portrait of Huxley; bibliography; index. 254 pp.

Baker, John R., and Jens-Peter Green. *Julian Huxley, Scientist and World Citizen, 1887-1975*. Paris: UNESCO, 1978.
Biography-bibliography. Julian Huxley was the first Director-General of the United Nations Educational, Scientific and Cultural Organization (UNESCO). The organization published this book to honor him. It has two parts. The first part contains a biographical sketch that summarizes the main features of Huxley's public life: his interest in birding and zoology, teaching, work for the Zoological Society, popular science writing, and work at UNESCO. The second part is a comprehensive bibliography of Huxley's writings and secondary sources about him. 184 pp.

Barthélemy-Madaule, Madeleine. *Lamarck the Mythical Precursor*. Cambridge, Mass.: MIT Press, 1982.
Biography-popular science. The author offers a study of the relations between science and ideology because, she says, scientists are increasingly uneasy over the orthodox view of inheritance of characteristics by genetic transmission alone. Inheritance of acquired characteristics—acquired from the environment—was first proposed by Jean Baptiste Pierre Antoine de Monet, Chevalier de Lamarck (1744-1829). She summarizes his life and career, explains his theory ("Lamarckism," almost always juxtaposed to Darwinism), and discusses the troubled history of his ideas after Charles Darwin published his findings on natural selection and geneticists linked evolution to genes. Although for general readers interested in science controversy, the book demands a basic knowledge of biology and science history. Bibliography; end notes; index. 174 pp.

Benson, Maxine. *Martha Maxwell: Rocky Mountain Naturalist*. Lincoln: University of Nebraska Press, 1986.
Biography. Martha Maxwell (1831-1881) gained fame at the Philadelphia Exposition in 1876 for her immense dioramas of animals from the Rocky Mountains. She shot and stuffed her

own exhibits, contributed specimens to the Smithsonian, and started her own Rocky Mountain Museum in Boulder, Colorado, in 1874. Benson recounts her daring life as a collector and naturalist but also devotes much of the book to Maxwell's close relationship with her daughter and her half sister as examples of female bonding. Written for a general readership. Photos; bibliography; end notes; index. 335 pp.

Berkeley, Edmund, and Dorothy Smith Berkeley. *Dr. Alexander Garden of Charles Town*. Chapel Hill: University of North Carolina Press, 1969.
Biography. A physician, Alexander Garden (1730-1791) was also a diligent naturalist who corresponded regularly with the other prominent naturalists of his day, such as Carl Linnaeus and John Ellis, yet Garden is little known. The authors try to revive Garden's reputation in their portrait of a man in love with collecting and studying the plants of South Carolina, where he immigrated in 1752. They trace him from Scotland to America, describe his medical practice and botanical and zoological research, and consider his place in the history of naturalists. Drawings; bibliography; footnotes; index. 379 pp.

Berkeley, Edmund, and Dorothy Smith Berkeley. *Dr. John Mitchell: The Man Who Made the Map of North America*. Chapel Hill: University of North Carolina Press, 1974.
Biography. Virginia-born John Mitchell (1711-1768) led a varied career as a physician whose contributions to botany prompted Carl Linnaeus to name a species of bird in his honor. As the Berkeleys explain, however, Mitchell also made discoveries in zoology, physiology, agriculture, and climatology; moreover, he also produced some of the best maps of North America during the eighteenth century. The authors argue that because Mitchell lived the last 22 years of his life in England, he was an important figure in English-colonial scientific and political affairs. Drawings; bibliography; footnotes; index. 283 pp.

Bibby, Cyril. *Scientist Extraordinary*. New York: St. Martin's Press, 1972.
Biography. In this volume, Bibby concentrates on the scientific

career of Thomas Henry Huxley; for his general biography of Huxley, see the entry below. Even in his own times, Huxley was best known as "Darwin's bulldog" for vigorously defending Charles Darwin's principle of natural selection against attacks by humanists and theologians. Huxley was also an eminent comparative anatomist and taxonomist, as Bibby shows. Huxley helped revolutionize the educational system of England and insisted that science be a part of it. As well as displaying Huxley's intellectual greatness, Bibby depicts the social milieu of nineteenth-century science in England. Photos and drawings; bibliography; index. 208 pp.

Bibby, Cyril. *T. H. Huxley, Scientist, Humanist and Educator*. New York: Horizon Press, 1960.
Biography. Bibby covers much of the same material as he does in his later book, *Scientist Extraordinary* (see the entry above), except that he devotes much more space to Thomas Huxley's personal life and family, work as an educator, and moral philosophy than to his scientific work. The book includes forewords by Huxley's two famous grandsons, Julian and Aldous. Photos and drawings; bibliography; end notes; index. 330 pp.

Blunt, Wilfrid. *The Compleat Naturalist*. London: Collins, 1971.
Biography. In this richly illustrated book for general readers, Blunt tells the story of Carl Linnaeus (1707-1778). Linnaeus was raised and educated in Sweden, explored Lapland, taught at the University of Upsala, was a royal physician, described Swedish flora, and devised an extremely influential classification system for plants during one of Sweden's most politically turbulent centuries. The many illustrations complement the text well in its explanations of Linnaeus' botanical work and its descriptions of the society of the times. Photos, portraits, drawings, and maps; bibliography; end notes; index. 256 pp.

Bonner, John Tyler. *Life Cycles: Reflections of an Evolutionary Biologist*. Princeton, N.J.: Princeton University Press, 1993.
Autobiography. With considerable wit and charm, and thorough clarity, Princeton professor John Tyler Bonner discusses his career and his guiding theme in biology: how the under-

standing of genetic expression and the mechanics of development serve evolution in the life cycles of organisms. His specialty is slime mold, as he puckishly claims in the book's first sentence. Neither the obscurity of that focus nor the details of biology he introduces in his story will repel general readers interested in science. He both describes the rise of modern biology and includes anecdotes about famous researchers. A perceptive, readable book for anyone who wants a glimpse into the world of modern biological sciences. Drawings and diagrams; bibliographical essay; index. 209 pp.

Bowlby, John. *Charles Darwin: A New Life*. New York: W. W. Norton, 1991.
Biography. A physician, Bowlby pursues a medical thesis: The chronic gastrointestinal illness that Charles Darwin (1809-1882) endured most of his life was primarily psychological. It came from an anxiety neurosis caused by his mother's early death and his father's critical attitude toward him. To escape the effects of the illness, Darwin devoted himself to his scientific labors, his evolution theory among them. Bowlby devotes the text to Darwin's family, friends, travels, colleagues, and health rather than to the scientific work itself. The book, in fact, presents a clearer view of Victorian society than of Victorian science. Many photos and drawings; annotated list of persons mentioned in the text; bibliographical essay and bibliography; end notes; index. 511 pp.

Boyd, J. Morton. *Fraser Darling's Islands*. Edinburgh, Scotland: Edinburgh University Press, 1986.
Biography-memoirs. The Scottish naturalist and biologist Fraser Darling (1903-1979) became famous before World War II for his popular accounts of biological research on Scottish islands. Boyd, his friend, recounts the heyday of Darling's research after a brief account of his background and education. The text quotes liberally from Darling's writings, published and unpublished, and the combination makes pleasant reading. It also explains lucidly the methods and results of Darling's work. Photos and diagrams; short bibliography; end notes; index. 252 pp.

Brodhead, Michael J. *A Soldier-Scientist in the American Southwest.* Tucson: Arizona Historical Society, 1973.

Biography. Brodhead relates the studies of birds and animals in the West by Elliott Coues (1842-1899), an Army surgeon who was a prolific writer of journal articles and produced three popular books that made him famous. The period covered in this pamphlet is primarily 1864-1865, but Brodhead prepares his intended readers, other historians, with a summary of Coues' background. Photos and artwork; end notes. 74 pp.

Brooks, John Langdon. *Just Before the Origin: Alfred Russel Wallace's Theory of Evolution.* New York: Columbia University Press, 1984.

Biography-history. The author's purpose in this book is to rescue from obscurity the ideas about organic change—that is, evolution—of Alfred Russel Wallace, an obscurity produced by the eclipsing figure of his elder contemporary, Charles Darwin. Much of the text is therefore explanation of Wallace's articles on the fauna of the Amazon and the Malay Archipelago. But Brooks depends upon much biographical information to explain the development, fine points, and significance of Wallace's biological work. There is almost no discussion of his background, education, or personal life. The intended audience appears to be college-educated general readers interested in science history. Reproductions of manuscripts and charts; bibliography; index. 284 pp.

Brooks, Paul. *The House of Life: Rachel Carson at Work.* Boston, Mass.: Houghton Mifflin, 1972.

Biography. Brooks devotes his book to the professional life of Rachel Carson by drawing on his own memories of her and those of colleagues, as well as her letters and publications. His purpose is to show how Carson researched and wrote her famous books about the sea, such as *The Sea Around Us* (1951), and her indictment of pesticides in *Silent Spring* (1962). As well as commentary, the book contains many excerpts from her writings. Photos and drawings; bibliography; end notes; index. 350 pp.

Browne, Janet. *Charles Darwin*. Vol. 1, *Voyaging*. New York: Alfred A. Knopf, 1995.
Biography. A book of high literary quality, this first of two volumes covers Darwin's life from his birth to his decision to write a book about the principles of natural selection (1809-1856). The story conveys both hominess and high adventure, as Darwin grew up on a rural estate in a close-knit family, studied in Edinburgh and Cambridge, sailed around the world on HMS *Beagle*, became a respected naturalist, married, and took up a country gentleman's life. Browne reveals the intensity of Darwin's intellect and his passion for collecting (as well as hunting) and explains the influences of older scientists, such as Robert Grant, John Stevens Henslow, and Charles Lyell, supplying mini-biographies of them as well as Darwin's famous father, Robert, and grandfather, Erasmus. She also explains Darwin's scientific discoveries and their historical context with unusual clarity and thoroughness. Photos and drawings; bibliography; end notes; index. 605 pp.

Cannon, H. Graham. *Lamarck and Modern Genetics*. Springfield, Ill.: Charles C Thomas, 1959.
Biography-popular science. A book with an axe to grind. In his somewhat bitter preface, Cannon accuses "orthodox geneticists" of ignoring his books, in which he defends the real ideas of Jean Baptiste Pierre Antoine de Monet, Chevalier de Lamarck from popular modern distortions. Cannon argues that Charles Darwin's contributions to evolution theory were either wrong or originated by others, such as Lamarck. This book briefly recounts the life of Lamarck and then explains Lamarck's theory of the inheritance of acquired characteristics and compares them to the tenets of Darwinism. End notes; index. 152 pp.

Cantwell, Robert. *Alexander Wilson, Naturalist and Pioneer*. Philadelphia, Pa.: J. B. Lippincott, 1961.
Biography. Alexander Wilson (1766-1813) was born to a smuggler in Scotland, left his homeland after he got into legal trouble, and became a teacher. He turned his attention to the birds of America and, although without formal scientific training, was for a long time the leading ornithological expert.

He wrote and illustrated a popular book, *American Ornithology*, and knew John James Audubon, Meriweather Lewis, and many other naturalists and explorers of the times. The table book format shows off the lovely details of Wilson's illustrations well, and the text, for general readers, depicts the difficult conditions early naturalists faced in collecting information and publishing their findings. Drawings; bibliography; index. 319 pp.

Carrighar, Sally. *Home to the Wilderness*. Boston, Mass.: Houghton Hifflin, 1973.
Autobiography. There is little science in this soul-searching look by Sally Carrighar at her troubled childhood and work in Hollywood and as a writer. When she discovered the joys of animal research in Sequoia National Park, she says, her life blossomed. Unfortunately, the book ends at that point, and one does not learn of her later zoological work in other parts of the world. Photos. 330 pp.

Clark, Eugenie. *Lady with a Spear*. New York: Harper and Brothers, 1951.
Autobiography. Eugenie Clark (b. 1922) won fame for her popular writings and Florida marine biology research station. This casual, delightful book looks back at her life before she settled at the station is an exotic story. She tells of the origin of her career in a trip to the New York Aquarium; education at Hunter College; work at the Scripps Institute of Oceanography; and research in the West Indies, Red Sea, and Pacific Islands, where she swam reefs in search of specimens to spear and study. An inspiring book for young readers considering a career in marine biology or those who enjoy tales about a first researcher to visit isolated areas. Photos and drawings. 243 pp.

Clark, Ronald W. *JBS: The Life and Work of J. B. S. Haldane*. New York: Coward-McCann, 1969.
Biography. John Burdon Sanderson Haldane (1892-1964) led a controversial, eccentric, energetic life, and Clark's shrewdly written appraisal of it evokes something of Haldane's irresistible force of character. In science, he helped pioneer the

mathematics of genetics, but his interests ranged far. He also contributed to physiology, biochemistry, diving technology, and mathematics. He wrote hundreds of popular science articles and books and served as Weldon Professor of Genetics at University College, London, until he resigned in disgust at age sixty-seven and emigrated to India, where he set up his own research institute. Short-tempered, quotably outspoken, and deeply humane, he seems, in Clark's capable treatment, a figure larger than life. Photos; bibliography; index. 326 pp.

Clark, Ronald W. *The Survival of Charles Darwin*. New York: Random House, 1984.
Biography-popular science. This masterfully written book discusses the deep conflict in Charles Darwin: hesitancy to destroy religious beliefs with his theory of evolution and ambition to make a name for himself in science. In the book's first section, Clark follows this conflict through Darwin's upbringing in a well-to-do upper-middle-class Victorian family, his uncertainties about a career, the voyage of the *Beagle*, and his career following publication of *The Origin of Species* (1859), when ambition triumphed. The second section traces the diffusion of the theory of natural selection, and the final section explains how it has changed with the discoveries of twentieth-century scientists, especially microbiologists. An entertaining, clear book for all readers interested in Darwin's life, Victorian science, or the development of the theory of evolution. Photos; bibliography; end notes; index. 449 pp.

Cohen, Michael P. *The Pathless Way: John Muir and American Wilderness*. Madison: University of Wisconsin Press, 1984.
Biographical interpretation. Cohen says this book does not attempt to retell the life of John Muir (1838-1914); instead, although grounded in biographical fact, it attempts to judge Muir as an old man by the standards and aspirations of his younger self. Thereby, Cohen traces the development of Muir's ideas about the wilderness. The book is best suited to readers interested in Muir. End notes; index. 408 pp.

Coleman, William. *Georges Cuvier, Zoologist*. Cambridge, Mass.: Harvard University Press, 1964.

Biography. After an opening chapter devoted to the family, youth, and education of Georges Cuvier (1769-1832), Coleman confines himself to Cuvier's work in animal anatomy and taxonomy at the Jardin des Plantes in Paris. Cuvier was a fierce opponent of the evolution of species, and Coleman relates how Cuvier retarded the development of a theory of evolution with his great influence in science. Coleman also explains how Cuvier helped set the definition of species on the basis of anatomical principles, work which won him lasting fame. Drawings; bibliographical essay; end notes; index. 212 pp.

Crick, Francis. *What Mad Pursuit*. New York: Basic Books, 1988.
Autobiography-popular science. Co-discoverer of the structure of deoxyribonucleic acid (DNA) with James Watson, Crick is among the founders of microbiology. In this book he explains how he got into the field and the problems that most attracted him: genetic replication, the genetic code, and how the brain works. He devotes most of the text to explications of molecular biology, in fact, and largely speaks of his own life and career in connection with it, although independently he describes his childhood, education, and research during World War II. Photos and diagrams; index. 152 pp.

Croker, Robert A. *Pioneer Ecologist: The Life and Work of Victor Ernest Shelford, 1877-1968*. Washington, D.C.: Smithsonian Institution Press, 1991.
Biography. Victor Ernest Shelford was among the first scientists to study animal habitats as interacting communities. As a professor at the University of Chicago and then the University of Illinois, he trained many of America's leading ecologists. He also helped found the Ecological Society of America and The Nature Conservancy. Croker wrote the book in order to study the origin of animal ecology as a discipline, in which Shelford was instrumental. For general readers interested in ecology and environmentalism. Photos; bibliography; end notes; index. 222 pp.

Cutright, Paul Russell, and Michael J. Brodhead. *Elliott Coues: Naturalist and Frontier Historian*. Urbana: University of Illinois Press, 1981.

Biography. Elliott Coues (1842-1899) was an Army surgeon who capitalized on the life in remote outposts by studying birds. His books on Colorado and Northwest birds made him an eminent ornithologist of his day. As the authors show, he also was a systematic zoologist, lexicographer, and historian, producing an incredible number of articles. They recount his youth in Washington, D.C., education at George Washington University and the Smithsonian Institution, Army service in the West, work on the Hayden survey of the forty-ninth parallel, and scientific and historical studies. Appendices list bird and mammal species that Coues first describes. Photos and drawings; large bibliography; end notes for each chapter; index. 509 pp.

Darwin, Francis. *The Life and Letters of Charles Darwin.* 2 vols. New York: Basic Books, 1959.
Autobiography-biography. A substantial introduction by George Gaylord Simpson reviews the origin of the main text and comments upon its style and content. The rest of the two volumes comprise a short biography by Francis Darwin, the autobiography of his father, Charles Darwin, and a selection of the elder Darwin's letters, all of which were first published in 1888. Drawings; bibliography; index.

Davies, John. *Douglas of the Forest.* Seattle: University of Washington Press, 1980.
Biography. Davies sandwiches long excerpts from the journals of David Douglas (1799-1834) between introductory chapters summarizing his life and describing his expedition up the Columbia River and a concluding chapter about his fame as a botanist. In his short life Douglas, a Scotsman, introduced dozens of Northwest species to the scientific world, including many of the region's trees, such as the lodgepole pine, Sitka spruce, and Douglas fir. Appendices of his botanical discoveries conclude the text. Photos and drawings; bibliography; index. 189 pp.

De Beer, Gavin. *Charles Darwin: Evolution by Natural Selection.* Garden City, N.Y.: Doubleday, 1963.
Biography. An evolutionary biologist and a former director of

the British Museum, de Beer surveys Charles Darwin's works reverently. He pays attention to the details of Darwin's life and the cultural milieu but emphasizes the scientific achievements and their broad impact on society. It is thus a thorough introduction to Darwin's career and to the theory of evolution for readers unfamiliar with either. Photos and drawings; bibliography; index. 290 pp.

Desmond, Adrian, and James Moore. *Darwin*. New York: Warner Books, 1991.
Biography. The authors offer a "social portrait" of Charles Darwin. Drawing from letters and notebooks first made available in the 1980's, they focus on the social milieu, especially Darwin's Whig sentiments, that surrounded him, and they "depict a man grappling with immensities in a society undergoing reform." Among the immensities, of course, is the theory of evolution, and the authors give it sufficient attention to support their thesis: Darwin's behavior and life-long gastrointestinal ailment resulted from worry and self-torment regarding his scientific findings. The closely argued text provides insight into the turbulent social reform movements of the times. Photos and drawings; bibliography; end notes; index. 808 pp.

Dibner, Bern. *Darwin of the Beagle*. New York: Blaisdell Publishing, 1964.
Biography. After a brief opening chapter on the family and youth of Charles Darwin, Dibner concentrates on the voyage of the *Beagle* around the world, during which Darwin acted as the captain's companion and naturalist. He describes in detail the peoples and fauna Darwin encountered, echoing Darwin's reaction in some cases, as in the author's apparent distaste for the inhabitants of Tierra del Fuego. The final quarter of the book discusses the theory of natural selection that Darwin formulated on the basis of his nature studies during the five-year voyage. Drawings; bibliography; index. 143 pp.

Dobell, Clifford. *Antony van Leeuwenhoek and His "Little Animals": Being Some Account of the Father of Protozoology and Bacteriology and His Multifarious Discoveries in These Disciplines*. New York:

Russell and Russell, 1958.
Biography-autobiography. This is a reprint of a 1932 publication in which Dobell supplies a long biographical essay about Anton van Leeuwenhoek (1632-1723) and then quotes at length from his letters and manuscripts about his observations of protozoa and bacteria. There is much valuable information in the book, but Dobell's style, unctuous and adulatory, cloys, and Leeuwenhoek's life and character get lost in the mass of details, some trivial, that Dobell lovingly discusses. Artwork and drawings; bibliography; index. 435 pp.

Dupree, A. Hunter. *Asa Gray: 1810-1888.* Cambridge, Mass.: Belknap Press of Harvard University, 1959. *Asa Gray, American Botanist, Friend of Darwin.* Baltimore, Md.: Johns Hopkins University Press, 1988.
Biography. Asa Gray was among the first scientists born and educated in the United States to earn an international reputation. He brought order and depth to the desultory collecting of most American botanists and perceived the similarity between New World and Old World plant species. He is best known as American's most vigorous defender of Darwinian evolution and the longtime intellectual foe of Louis Agassiz. Dupree's scholarly biography, pleasurable reading for anyone interested in science, thus affords the drama of a grand debate, a view of American science during its youth, and a portrait of a remarkably determined and resourceful intellect. Photos; end notes; index. 503 pp.

Evans, J. Edward. *Charles Darwin: Revolutionary Biologist.* Minneapolis, Minn.: Lerner Publications, 1993.
Biography for readers 12 years and older. Evans introduces Charles Darwin's theory of evolution, announced in 1858, not as the first theory of evolution; instead, he says, it was the first data-based consideration of a topic that other scientists had been exploring for the previous fifty years. Then Evans relates Darwin's life and the character traits that led him, reluctantly, to his theory. The skillful writing style and many illustrations make this an excellent book for young readers coming to Darwin and the theory of evolution for the first time. Photos and drawings; bibliography; index. 111 pp.

Fagin, N. Bryllion. *William Bartram, Interpreter of the American Landscape*. Baltimore, Md.: Johns Hopkins University Press, 1933.

Biography. According to Fagin, the *Travels* by William Bartram (1739-1823) became a classic in nature writing. It was a book in the Romantic style (Samuel Taylor Coleridge and Thomas Carlyle were among its enthusiastic admirers), and it influenced several generations of readers, forming their views of the wilderness. Fagin studies the extent and nature of Bartram's influence in this book for literary scholars, but he also reviews Bartram's life and career as a naturalist, and botanist in particular. Accordingly, chapters address his life, character, and philosophy of nature; his landscape art; and the use of his ideas by later writers. Bibliography; index. 229 pp.

Fellows, Otis E., and Stephen F. Milliken. *Buffon*. New York: Twayne Publishers, 1972.

Biography. Georges-Louis Leclerc, comte de Buffon (1707-1788) was a star of French science during the Enlightenment for his mammoth *Histoire Naturelle* (1749-1788), which attempted to catalog and analyze the fauna then known. The authors discuss Buffon's varied scientific and literary accomplishments in chapters devoted to his contemporary reputation, youth, rise to fame, and scientific work on the life force. In lucid prose, they ponder whether his work foreshadows that of Charles Darwin and take pains to place Buffon's ideas in the context of eighteenth-century France. Bibliography; end notes; index. 186 pp.

Fichman, Martin. *Alfred Russel Wallace*. Boston, Mass.: Twayne Publishers, 1981.

Biography. Fichman begins with a chapter-long summary of the life and career of Alfred Russel Wallace (1823-1913). Following chapters take up his theory of natural selection, biogeography, human evolution, and liberal social and political views. Fichman also examines Wallace's intense interest in spiritualism and mesmerism. A succinct, plainly written summary, especially suited to college students. Photo of Wallace; bibliography; end notes; index. 188 pp.

Fischer, Ernst Peter, and Carol Lipson. *Thinking About Science: Max Delbrück and the Origins of Molecular Biology*. New York: W. W. Norton, 1988.
Biography. A 1969 Nobelist for his contributions to molecular biology, Max Delbrück (1906-1981) set out to be an astronomer, turned to quantum physics, and, inspired by the complementarity principle of Niels Bohr, became a biophysicist. He grew up in an illustrious Berlin academic family but became an expatriate in the United States because of the Nazi regime. The authors tell his story in great detail, drawing on unpublished autobiographical accounts, records, and his many publications, as well as reminiscences of his friends and colleagues. The narrative is loose-jointed and sometimes awkwardly written; in general the long sections explaining Delbrück's research and discoveries are more fluent than the sections about his life and politics. Photos; bibliography; index. 334 pp.

Ford, Alice. *John James Audubon*. Norman: University of Oklahoma Press, 1964.
Biography. Dramatically written, this book portrays John James Audubon as a romantic, emotional, vastly energetic figure. Ford recounts his birth in the Caribbean to a French sea captain and his Creole mistress, youth in France and the United States, wanderings through the American woods, developing skill with painting, trips to England, and publication of *Birds of America*, *Ornithological Biography*, and *Viviparous Quadrupeds of North America*. As well as Audubon's verve, Ford also shows his philosophical side and his dealings with his family. Photos and artwork; register of Audubon's paintings; bibliography; footnotes; index. 488 pp.

Ford, Brian J. *Single Lens*. New York: Harper and Row, 1985.
Biography-history. Ford recounts the history of the simple microscope, which has a single lens instead of the objective lens and eyepiece of most modern microscopes. Because Anton van Leeuwenhoek (1632-1723) constructed them and used them to discover cells, nuclei, and bacteria, Ford naturally discusses the Dutch scientist and his work at length. Although the instrument seems a narrow scholarly focus for

popular science, the book is aimed at a general audience. Photos and drawings; bibliography; indexes. 182 pp.

Ford, Corey. *Where the Sea Breaks Its Back.* Boston, Mass.: Little, Brown, 1966.
Biography-history. Ford tells the story of George Wilhelm Steller (1709-1746) and his study of Alaskan fauna and flora while serving as the physician on Vitus Bering's exploration of the area in 1741. The account makes Steller seems a very remarkable man and is written to emphasize the danger and excitement of the voyage. Steller first described many bird and fish species that are now well known, including Steller's jay, and some mysteries, such as Steller's sea monkey, never seen by anyone else. Written to captivate a popular audience, the book is not a comprehensive biography; Corey says little about Steller's background so that he can keep his narrative tuned to adventures in early Alaska. Drawings; bibliography. 206 pp.

Frick, George Frederick, and Raymond Phineas Stearns. *Mark Catesby, the Colonial Audubon.* Urbana: University of Illinois Press, 1961.
Biography. A pioneering scientific illustrator, Mark Catesby (1683-1749) was the most productive gatherer of plant and animal specimens in colonial America. His publications afforded later scientists, such as Carl Linnaeus, crucial evidence for biological taxonomy and theories. His style of illustration was widely admired and imitated; among his imitators was John James Audubon. Nonetheless, he has become an obscure figure in the history of zoology, and the authors intend to revive Catesby's reputation in their book. To do so, they provide a moderately detailed account of his youth, education, and career and analyze his contributions in the context of eighteenth-century zoological and botanical thought. They also supply many impressive examples of Catesby's illustrations. For readers familiar with the history of biology. Drawings; bibliographical essay; footnotes; index. 137 pp.

Frisch, Karl von. *A Biologist Remembers.* Oxford, England: Pergamon Press, 1967.

Autobiography. Karl von Frisch (1886-1982) became famous for his studies of bee behavior, especially how they communicate through dances to locate food sources. For his work on animal behavior he shared the 1973 Nobel Prize for Physiology and Medicine with Konrad Lorenz and Nikolaas Tinbergen. Here he recalls his family background, education, and career, spent mostly at the Zoological Institute in Munich, up until 1950. He also recounts two research trips to the United States and his experiences during World War II, when his mixed racial heritage caused him trouble with the Nazi authorities. He carefully explains his methods for studying bees and the results. Photos and drawings; bibliography; index. 200 pp.

Galdikas, Biruté. *Reflections of Eden: My Years with the Orangutans of Borneo*. Boston, Mass.: Little, Brown, 1995.
Autobiography. Like Jane Goodall and Dian Fossy, Biruté Galdikas (b. 1946) lived for years in the wilderness studying great apes at first hand under the sponsorship of Louis Leakey. In her case the research tracked orangutans in Borneo, apes that were largely a mystery, or misunderstood, before Galdikas took up her station. She writes of her research, which began early in the 1970's, with drama and lyricism. The tone is pensive, always engaging, whether she is describing the behavior of the orangutans that bonded with her or explaining the history and technicalities of primate research. This is an inspiring book for anyone interested in field research. Photos. 408 pp.

George, Wilma. *Biologist Philosopher*. London: Abelard-Schuman, 1964.
Biography. The author assesses the theories of Alfred Russel Wallace in the history of biology. George passes quickly over his upbringing and education and then examines in detail Wallace's specimen-gathering expeditions to the Amazon and the Malay Archipelago. From these experiences Wallace formulated, independently of Charles Darwin, the theory of natural selection and protective mimicry, but, George insists, as great an accomplishment was Wallace's study of the distribution of species, ancient and modern. This biogeography

enabled him to demarcate regions of independent development of species, a milestone in evolutionary theory. George also discusses Wallace's socialistic political views—Victorian capitalism repelled him—and his interest in spiritualism. An engrossing story of the man who stood in Darwin's shadow yet was as brilliant, eccentric, versatile, and eloquent about his discoveries. For a general audience. Photos and drawings; bibliography; index. 320 pp.

Goldberg, Jake. *Rachel Carson*. New York: Chelsea House, 1992.
Biography for readers eight years and older. Goldberg relates Rachel Carson's fascination with the oceans and her crusade to expose the dangers of insecticides, especially DDT. He tells young readers that these interests and her widely popular books earned her the reputation as the founder of modern ecology. Photos; glossary. 79 pp.

Goldschmidt, Richard B. *In and Out of the Ivory Tower*. Seattle: University of Seattle Press, 1960.
Autobiography. Richard B. Goldschmidt (1878-1958) was a German-born biologist specializing in genetics, the physiology of sex, and the geographical variation of species, but these reminescences have little in them about the details of his research, although he summarizes his scientific interests in an appendix. Instead, in the main narrative he recounts his extensive travels around the world, art collecting, stay in Japan, and tenure at the Kaiser Wilhelm Institute in Berlin. He immigrated to the United States to escape Nazi persecution of Jews and became a professor at the University of California in Berkeley. Often charmingly related, his recollections include exotic adventures and the politics of staid pre-World War II European academia. Photos; bibliography; index. 352 pp.

Goodall, Jane. *Through a Window: My Thirty Years with the Chimpanzees of Gombe*. Boston, Mass.: Houghton Mifflin, 1990.
Memoirs. Jane Goodall (b. 1934) is among the most famous animal behavior researchers of the twentieth century. Her books and documentary films inspired a generation of new scientists to follow her adventurous lead. Here, Goodall's richly descriptive prose limns a vivid picture of her life in

Zaire, where she had to weather a war, but most of the book recounts the social and individual traits of the chimpanzees she has studied. A thoughtful, engrossing book for all readers interested in zoology and field research. Photos; index. 268 pp.

Graustein, Jeannette E. *Thomas Nuttall, Naturalist: Explorations in America 1808-1841*. Cambridge, Mass.: Harvard University Press, 1967.
Biography. The author confines her account to the 33 years that Thomas Nuttall (1786-1859) lived in America. He explored from the Columbia River valley and California through the Midwest and Great Lakes to the Appalachians and Atlantic seaboard, collecting plant specimens and studying land formations as he went. He compiled the first taxonomic botanical work of broad scope for the nation and published the first popular guide to North American birds. Graustein thoroughly describes Nuttall's wilderness adventures, as well as his years at the Harvard botanical garden. Her aim, she says, is to correct the many factual errors that overly enthusiastic popular writers had made about Nuttall's life and to show his great influence on American naturalists before family duties forced him to return to England. Photos and drawings; brief bibliography; end notes; index. 481 pp.

Gruber, Howard E. *Darwin on Man*. 2d ed. Chicago, Ill.: University of Chicago Press, 1981.
Biography-psychological analysis. The author applies psychological theory to explain Charles Darwin's thinking as he struggled to synthesize the theory of evolution. Accordingly, Gruber considers only a narrow slice of Darwin's life following the voyage of the *Beagle* (1831-1836). He relies heavily on Darwin's notebooks from 1837 to 1839. The intended audience is other Darwin scholars. Drawings; footnotes; index. 310 pp.

Gunther, Albert E. *A Century of Zoology at the British Museum*. Folkstone, England: William Dawson and Sons, 1975.
Biography-history. Gunther relates the zoological studies at the British Museum from 1815 to 1914 by profiling the two Keepers of Zoology during the period. Both were leading researchers of their day. John Edward Gray (1800-1875) was a

botanist and taxonomist. Albert Gunther (1830-1914, the author's grandfather) was a physician who turned to the study of fish and birds. Gunther writes primarily for other scholars. Photos and drawings; bibliography; index. 533 pp.

Hagberg, Knut. *Carl Linnæus*. New York: E. P. Dutton, 1953.
Biography. For general readers, Hagberg recounts the life and career of the great Swedish botanist and taxonomist, Carl Linnaeus (Carl von Linné, 1707-1778). He describes in detail Linnaeus' trip to Lapland and struggle to win a professorship at the University of Upsala, but the main focus of the text is on his studies of Swedish plant life and his system for classifying plants. Hagberg also explains Linnaeus' philosophy and considers his influence on theories concerning the origin of species. Photos and drawings. 264 pp.

Hamerstrom, Frances. *My Double Life: Memoirs of a Naturalist*. Madison: University of Wisconsin Press, 1994.
Autobiography. A student of Aldo Leopold, Frances Hamerstrom (b. 1907) was an ornithologist who studied prairie chickens, hawks, and owls. She worked to protect endangered species and wrote popular books about wildlife. This book consists of short glimpses of her youth—she makes herself appear to be a thoroughly spoiled child and hellion—in a rich family, education, and career. Often ironical and humorous, the text nevertheless describes her pioneering birding research well for general readers. Photos and drawings. 316 pp.

Harvey, Athelston George. *Douglas of the Fir*. Cambridge, Mass.: Harvard University Press, 1947.
Biography. Harvey presents the fruits of years of research in the life of David Douglas, undertaken independently simply because Douglas intrigued him. Known in the Northwest because the Douglas fir is named after him, Douglas, a botanist, first described to other scientists many North American species. These he found during his explorations of the Columbia River and Puget Sound. He also studied the flora of the Hawaiian Islands. Portrait of Douglas; bibliography; footnotes; index. 290 pp.

Hayes, Harold T. P. *The Dark Romance of Dian Fossey*. New York: Simon and Schuster, 1990.

Biography. With the help of Louis Leakey, Dian Fossey left her job as a physical therapist in for crippled children Kentucky and moved to Rwanda to study mountain gorillas in their natural habitat. Her careful observations of their behavior advanced knowledge about these shy primates a great deal, and her militant efforts to protect them from poachers led to her murder in 1985. To many she became a martyr in the conflict between those who want to understand nature and those who want to exploit it. The exotic setting in her mountain jungle research camp, her troubles with local authorities, her single-minded determination to study the gorillas, and her self-reliance are all elements of romance in Hayes' telling, and the sinister military figures and poachers are the dark side of that romance. The book plays up these elements for a popular readership and brings in Fossey's scientific achievements only as backdrop. Photos; bibliography; index. 351 pp.

Heinrich, Bernd. *In a Patch of Fireweed*. Cambridge, Mass.: Harvard University Press, 1984.

Autobiography-popular science. Heinrich Bernd (b. 1940) writes to give general readers some feeling of the adventure and fun involved in biological research, qualities that are usually lost in the published results of such research. His first three chapters are directly autobiographical, but only insofar as they show his connection with biology. The remaining twelve chapters recount his research in the bees, beetles, caterpillars, and other insects of North America, Africa, and New Guinea. For students interested in field research or any reader who enjoys vicarious adventure, this book provides both entertainment and insight into the life of a biologist. Drawings. 194 pp.

Herbst, Josephine. *New Green World*. New York: Hastings House, 1954.

Biography. Herbst describes the lives of John Bartram (1699-1777) and his son William Bartram (1739-1823) in loving detail. To her these men represent a heroic but vanished breed who lived with great zest and immediacy in nature, observing, treasuring, and understanding the value and context of plants. The elder Bartram was a Pennsylvania farmer and amateur botanist; his son became a wandering artist and collector. Their stories reveal a time before botany in the United States became an organized scientific discipline, a time when everywhere an observant person went there were new species to discover. Drawings; bibliography. 272 pp.

Herrick, Francis Hobart. *Audubon the Naturalist.* 2 vols. New York: D. Appleton-Century, 1938; Dover, 1968.
Biography. The original version of this book appeared in 1917, and it was the first biography to tell of John James Audubon's family background and birth, hitherto guarded family secrets. The first volume concerns these matters, as well as Audubon's education in France and America, failure at business, skill in art, and first trips in America searching for new bird species. The second volume covers his explorations of Florida and the Eastern seaboard, major publications, and surviving family members. Photos and drawings; bibliography; footnotes; index in Vol. 2.

Hoagland, Mahlon. *Toward the Habit of Truth: A Life in Science.* New York: W. W. Norton, 1990.
Autobiography. In this ably written book, Hoagland (b. 1921) ponders the influences that led him into biochemistry and on to his discovery of transfer ribonucleic acid (tRNA). A big influence was his father, Hudson Hoagland, co-founder of the Worcester Foundation for Experimental Biology with Gregory Pincus, who developed the first oral contraceptive pill. Hoagland eventually took over leadership of the foundation from his father, but before that, during his years of biochemical research, he came to know many of the field's luminaries, such as Frederick Sanger and Francis Crick. His is the story of a model scientific career in mid-century America. Photos; end notes; index. 206 pp.

Hutchinson, G. Evelyn. *The Kindly Fruits of the Earth.* New Haven, Conn.: Yale University Press, 1979.
Autobiography. George Evelyn Hutchinson (b. 1903), a

British-born Yale zoologist, wrote these reminiscences because younger colleagues wanted to know what it was like to learn biology in the first quarter of the twentieth century. He obliges them by relating his youth and education in an English public school and Cambridge in great detail for more than half of the book, describing his teachers and fellow students. The final two chapters concern his teaching career, first in South Africa and then at Yale. Little of the text addresses zoological research. Photos and drawings; index. 264 pp.

Huxley, Julian. *Memories*. 2 vols. New York: Harper and Row, 1970-1973.
Autobiography. A widely read popular writer, teacher, and researcher, Julian Huxley (1887-1975) writes of his life with candor and old-world charm. He made his name in science as an expert on animal behavior, particularly that of birds, and his reputation was so great that he eventually became the first director of the United Nations Educational Scientific and Cultural Organization (UNESCO). These two volumes, often witty and sometimes poignant, detail the life of an upper-class English intellectual and show readers the family life of Huxley's many famous relatives, such as his grandfather Thomas H. Huxley and his brother Aldous Huxley. There is much information about other scientists as well, for Huxley was long a central figure in the biological sciences. Photos; indexes.

Huxley, Julian, and H. B. D. Kettlewell. *Charles Darwin and His World*. New York: Viking, 1965.
Biography. In this short, vividly illustrated account, the authors consider three phases in Charles Darwin's life: his youth, the intellectual transformation caused by his voyage around the world in the *Beagle* , and his illness-plagued years writing and then defending *The Origin of Species* (1859), as well as conducting other biological studies. Only by studying these phases, the authors argue, can readers understand how deeply Darwin affected the concepts of humanity, science, and religion. Although the narrative is brief, it is a gracefully written, thoughtful introduction for general readers who have

read nothing about Darwin before. Photos and drawings; index. 144 pp.

Huxley, Leonard. *Life and Letters of Sir Joseph Dalton Hooker*. 2 vols. London: John Murray, 1918.
Biography. The best friend of the botanist Joseph Hooker (1817-1911) was Thomas Huxley, and it is Huxley's son Leonard who wrote this detailed biography. Having access to Hooker's correspondence, which included letters to Charles Darwin and other great scientists of the day, Huxley supplements his personal knowledge of Hooker with abundant excerpts from the letters. The text follows Hooker to the Antarctic, Tasmania, India, Palestine, America, and the Himalayan Mountains. Huxley explains not only Hooker's botanical research but also his thoughts on evolution, Buddhism, and geology. For readers interested in botany and exploration, this is an exciting book about an undaunted traveler and brilliant scientist. Photos and drawings; bibliography; footnotes; index.

Huxley, Leonard. *Life and Letters of Thomas Henry Huxley*. 2 vols. New York: Appleton, 1900.
Biography. Leonard Huxley memorializes his father, telling his life story in a collection of letters linked by narrative. The intention, according to the author, is not to present the elder Huxley's scientific achievements, philosophical ideas, or educational efforts in technical detail; rather, it is to provide a "picture of the man himself"—that is to say, his character and temperament and the circumstances in which he worked. The author tries to keep the narrative impersonal; nevertheless, this biography is an invaluable source of information about the elder Huxley's personal life and his relationship with his large, talented family. Photos and drawings; bibliography; index.

Iltis, Hugo. *Life of Mendel*. New York: W. W. Norton, 1932.
Biography. First published in 1924 in Germany, this book was an early recognition of Gregor Mendel's importance as a theorist. Iltis not only relates the life story and scientific work of Mendel in generous detail; he also describes how the monk-

biologist's work sank into obscurity and was rediscovered early in the twentieth century, just when genetics was emerging as a biological discipline. A thoroughly researched, lucid account, still a good source for general readers interested in the history of biology. Photos, drawings, and diagrams; footnotes; index. 336 pp.

Jensen, J. Vernon. *Thomas Henry Huxley: Communicating for Science*. Newark: University of Delaware Press, 1991.
Biography. The author calls this a "rhetorical biography" of Thomas Henry Huxley because the focus is on Huxley's oratorical skill and the occasions he used it most effectively. The opening chapter offers an overview of Huxley's life and career, and the first chapter concerns his early confidantes, that is, three women with whom he discussed and debated his ideas, honing his powers of expression. Subsequent chapters regard his first public address, famous debates—especially the one with Bishop Wilberforce when Huxley defended Darwinism—theological speeches, conversations with friends, and rhetorical legacy. Photo of Huxley; bibliography; end notes; index. 253 pp.

Johnson, Osa. *I Married Adventure*. Philadelphia, Pa.: J. B. Lippincott, 1940.
Autobiography. An entertaining bit of early twentieth century Americana. Osa and Martin Johnson, from small-town U.S.A., became photographers, documentary movie makers, and specimen collectors for the American Museum of Natural History during their life of safaris in Africa. They were world famous, having hobnobbed with celebrities such as Charlie Chaplin and the Duke of York, when a plane crash killed Martin and left Osa badly injured. Not much science is involved in their story, but the book vividly depicts the life of Jazz Age naturalists who were really big game hunters and entrepreneurs. Photos; index. 376 pp.

Jordanova, L. J. *Lamarck*. Oxford, England: Oxford University Press, 1984.
Biography. A brief, lucid, handy guide to the life and ideas of Jean-Baptiste Pierre Antoine de Monet de Lamarck. Jordanova

argues that Lamarck's theory of inherited traits is now considered simply wrong, but that does not mean students of science should neglect the theory because it formed a principal part of the debate that led to modern evolution theory. Accordingly, Jordanova explains the genesis and tenets of Lamarckism. Lamarck's research in chemistry and meteorology, as well as his Enlightenment-influenced scientific philosophy, also receives attention. Short bibliographical essay; index. 118 pp.

Kamen, Martin D. *Radiant Science, Dark Politics: A Memoir of the Nuclear Age*. Berkeley: University of California Press, 1985.
Autobiography. A cleverly, accurately titled book. Martin D. Kamen (b. 1913) co-discovered carbon 14, which he used as a radioactive tracer to study photosynthesis. He also worked with E. O. Lawrence at his radiation laboratory and as a participant in the Manhattan Project knew many of the most famous scientists and chemists of modern science, including Linus Pauling, J. Robert Oppenheimer, and Niels Bohr. He used radioactive isotopes in biological experiments, the "radiant" of the title. During the Cold War he was falsely accused of spying and scrutinized by the House Committee on Un-American Activities. He cleared his name only after great struggles in the "dark" world of red-scare politics. His story reveals the ambivalent nature of government-supported big science projects. For general readers, although some knowledge of the history of twentieth-century physics is helpful. Photos; end notes; index. 348.

Keim, Charles J. *Aghvook, White Eskimo*. College: University of Alaska Press, 1969.
Biography. Keim tells the story of a determined, largely self-taught naturalist, Otto William Geist (1888-1963), who became a University of Alaska scientist. German-born, Geist came to North America after World War I and collected plant and animal specimens, as well as mammoth fossils, in the state of Alaska and surrounding waters, but most remarkably, he was adopted by Eskimos and studied their culture closely. An adventurous life story related by a colleague. Photos; index. 313 pp.

Keller, Evelyn Fox. *A Feeling for the Organism*. New York: W. H. Freeman, 1983.

Biography. This remarkably insightful book not only brings to life the mind and philosophy of one of America's great scientists, Barbara McClinton (1902-1992); it also suggests that the standard reductionist view of the scientific method has shortcomings. A geneticist and cytologist, McClintock won the 1983 Nobel Prize in Physiology or Medicine for discovering genetic transposition. As Keller tells it, the story linking the discovery, made in the 1940's, to recognition of its meaning and worth, in the 1970's, is dramatic and revelatory. Although having a world reputation for other discoveries in genetics (working on maize), McClintock was scorned by her colleagues for her ideas about transposition, her intuitional approach to research, and her contention that a species' genome responds to environmental stress by regulating mutations. This hypothesis ran counter to the traditional view that evolution and genetic variation proceed randomly. McClintock remained a loner in science most of her life, first because she was a woman in a man's world and second because of her revolutionary ideas. The ideas triumphed, as did, finally, McClintock, a fact which illustrates, as Keller puts it, "the function of dissent in science." A well-written account, the book contains explanations of genetics and cell division that are sophisticated and lucid. Photos and drawings; end notes; glossary; index. 235 pp.

Kendall, Martha E. *John James Audubon: Artist of the Wild*. Brookfield, Conn.: Millbrook Press, 1993.

Biography for readers eight years and older. America's best-known naturalist-artist, Audubon learned the habits of birds as well as recorded them in paintings. Kendall summarizes his life and explorations, and the illustrations beautifully portray Audubon's work and his era. Photos and artwork; short bibliography; index. 48 pp.

Kessel, Edward Luther. *Autobiographical Anecdotes (I Was a Preacher's Kid)*. South San Francisco, Calif.: Insect Associates, 1989.

Autobiography. The publishers issued this volume as a tribute to Edward Luther Kessel (b. 1904). It contains his recollections of his boyhood, education, travels, and career as an entomologist at the University of San Francisco, especially the study of fire bugs and flies. An article about balloon flies by Kessel accompanies the narrative. Photos; bibliography. 234 pp.

King-Hele, Desmond. *Doctor of Revolution: The Life and Genius of Erasmus Darwin*. London: Faber and Faber, 1977.
Biography. Erasmus Darwin (1731-1802), Charles Darwin's grandfather, was an eminent physician and promoter of the Industrial Revolution. As King-Hele shows, Darwin also contributed to the understanding of plant nutrition and photosynthesis, writing an influential long poem, *The Botanic Garden* (1791), on such matters. He also explained the main process of cloud formation. Sociable and witty, Darwin knew many of the leading intellectuals of the day, including James Watts, Benjamin Franklin, and Joseph Priestley. King-Hele provides a vivid portrait of a man who exemplifies eighteenth-century English intellectual life. Photos, drawings, and diagrams; bibliography; end notes; index. 361 pp.

Kolankiewicz, Leon. *Where Salmon Come to Die: An Autumn on Alaska's Raincoast*. Boulder, Colo.: Pruett Publishing, 1993.
Memoirs. A pleasantly written book about a research project on coho salmon. Kolankiewicz occupied a remote island camp in southeast Alaska for the project, encountering strange coworkers and grizzly bears as well as coho. The book consists of letters home and journal entries and contains much information on salmon, as well as insight into government wildlife research. Photos. 126 pp.

Konnyu, Leslie. *John Xantus, Hungarian Geographer in America (1851-64)*. Cologne, Germany: American Hungarian Publisher, 1965.
Biography. Although John Xantus is best known as a naturalist in general, and ornithologist in particular, Konnyu insists that his geographical work in America as cartographer for the Pacific Railroad Company and his writings in Hungarian about the exploration of the West made him a leading interpreter of America to Europeans. This is an odd little book but

of interest to historians of science and of the West. Photos and drawings; bibliography; footnotes. 48 pp.

Kudlinski, Kathleen V. *Rachel Carson, Pioneer of Ecology*. New York: Viking Kestrel, 1988.
Biography for readers ten years and older. The author opens with Rachel Carson, age seven, listening to a conch shell, an incident which started her love for the sea, and readers find out how a determined person can overcome obstacles to pursue such a love. Kudlinski writes of Carson's childhood, education, and career as a civil servant but focuses most on the writing and aftermath of *The Sea Around Us* (1950) and *Silent Spring* (1963). Drawings. 55 pp.

Lear, Linda. *Rachel Carson: Witness for Nature*. New York: Henry Holt, 1997.
Biography. Far more thorough in its research than previous books about Rachel Carson, this biography discusses her family life in great detail and examines the composition and reception of her books, especially *The Sea Around Us* (1951) and *Silent Spring* (1962). Lear interviewed friends, family, and colleagues of Carson and combed her manuscripts and letters. Her principle thesis is that Carson had a "ferocious will" coupled with a deep love of nature. Both, Lear intimates, she got from her domineering mother. Her rhythmic, image-laden writing style and exacting research skills she developed largely on her own, although with key help from mentors in college. Emphasizing Carson's courage and hard work, Lear paints her almost as a martyr for art and science, beset at home with family problems and attacked by the "scientific establishment" for her environmental views and crusade against pesticides. A readable examination of one of the most influential science writers of the twentieth century, the book is especially valuable for its analysis of Carson's writing habits. Photos; bibliography; end notes; index. 634 pp.

Leighton, Gerald. *Huxley: His Life and Work*. London: T.C. and E.C. Jack, 1912.
Biography. An early and very sketchy popular life of Thomas Henry Huxley. The first three chapters—about one-third of

the book—lay out Huxley's life and career in a particularly stuffy style. Following chapters summarize his scientific works, popularization of science, educational views, and reputation among contemporaries. Bibliography. 94 pp.

Levi-Montalcini, Rita. *In Praise of Imperfection*. New York: Basic Books, 1988.
Autobiography. Eloquently, elegantly written memoirs by the winner of the 1986 Nobel Prize for Medicine or Physiology. Rita Levi-Montalcini (b. 1909) grew up in Turin, Italy, and trained as a doctor. Because she was Jewish, she had to hide out from the fascist government, and later the Nazis, in Italy. During this time, she nevertheless performed experiments on chicken embryos in a tiny closet, work that gave her experience for later experiments in the United States during which, with Stanley Cohen, she discovered nerve growth factor and its role in cell differentiation. Hers is an extraordinary story, both personally and scientifically, which she delivers to general readers with clarity and broad cultural sophistication. Photos; end notes; index. 220 pp.

Lincecum, Jerry Bryan, and Edward Hake Phillips, eds. *Adventures of a Frontier Naturalist: The Life and Times of Dr. Gideon Lincecum*. College Station: Texas A&M University Press, 1994.
Autobiography-biography. Gideon Lincecum (1793-1874) was a physician in frontier Texas who became an authority on local plants, especially their medicinal value. One of this book's editors, Jerry Lincecum, is a direct descendant of the doctor and provides a biographical sketch so that readers will understand the historical and scientific context of his forebear. The rest of the text is a composite of four memoirs that Gideon Lincecum wrote about his life and botanical pursuits in Texas. Photos; bibliography; end notes; index. 321 pp.

Luria, Salvador E. *A Slot Machine, a Broken Test Tube*. New York: Harper and Row, 1984.
Autobiography. Salvador Luria (1912-1991) shared the 1969 Nobel Prize for Physiology or Medicine for his pioneering studies of bacteriophages. Long before that he had a wide

influence on molecular biology, training other Nobel laureates and establishing a center of biochemical and cancer research at the Massachusetts Institute of Technology. Here Luria ruminates on his native Italy, which he fled to escape Fascist persecution of Jews, his academic career in the United States, and his colleagues. He arranges the book by subjects, rather than chronologically, so there is some repetition: his youth; science; teaching; use of the imagination; politics; and the influence of emotions. Although he sometimes lapses into bland praise of colleague after colleague, he writes adroitly for the most part, and the passages explaining his socialist views are provocative reading. Although even the passages about biophysics and molecular biology are for general readers, the book's cultivated tone and cultural sophistication make it an intellectual's fare. Photos and diagrams; index. 228 pp.

Lwoff, André, and Agnes Ullmann, eds. *Origins of Molecular Biology: A Tribute to Jacques Monod*. New York: Academic Press, 1979.

Biography and reminiscences. One of the editors of this volume, Lwoff, shared the 1965 Nobel Prize for Physiology or Medicine with Jacques Monod (1910-1976) for their research into how genes regulate reactions between cells. This volume contains a biographical sketch of Monod's life and career, and then 31 short essays by colleagues and students highlight episodes of his research or private life, although the emphasis is upon him as a scientist. Among the authors are leading biologists of the twentieth century, such as Salvador Luria and Francis Crick. The essays sometimes involve explanations of biological processes that assume an advanced knowledge in readers. Photos and drawings. 246 pp.

McCay, Mary A. *Rachel Carson*. New York: Twayne Publishers, 1993.

Biography. To McCay, the last book by Rachel Carson, *Silent Spring*, continued Carson's crusade to teach people about the natural world, the sea especially, and respect the interconnectedness of its creatures. *Silent Spring* was intended to show that pesticides endangered the precarious balance of life. McCay develops these themes through chapters about

Carson's youth and education, major publications, research on pesticides, and place in the tradition of American naturalists and environmentalists. A compact, informative book for general readers interested in ecology and nature writing. Photo of Carson; bibliography; end notes; index. 122 pp.

McGovern, Ann. *Shark Lady: True Adventures of Eugenie Clark*. New York: Four Winds Press, 1978.
Biography for readers ten years and older. Eugenie Clark grew from a child fascinated by fish in a New York City aquarium to a professional ichthyologist running an aquarium of her own—actually, the Cape Haze Marine Laboratory—in the 1950's. She also studied South Seas fish and reefs before World War II and became a world expert on sharks. McGovern tells Clark's story with humor and drama, intending it to inspire her young readers. Drawings; short bibliography. 83 pp.

McKinney, H. Lewis. *Wallace and Natural Selection*. New Haven, Conn.: Yale University Press, 1972.
Biography. McKinney writes to clear up a central question about Alfred Russel Wallace: Just when and how did he recognize that natural selection lies behind the evolution of life? The book therefore does not follow Wallace's entire life, only the period 1823-1858, covering his development as a naturalist to his first public announcement, with Charles Darwin, of natural selection. McKinney discusses Wallace's trip to the Amazon, publications, stay in the Malay Archipelago, study of Thomas Malthus' population theory, influence on Darwin and Charles Lyell, and place in evolutionary biology. Reproductions of manuscripts and drawings; bibliography; index. 193 pp.

McNeil, Maureen. *Under the Banner of Science: Erasmus Darwin and His Age*. Manchester, England: Manchester University Press, 1987.
Biography. Charles Darwin's grandfather, Erasmus Darwin, published poems describing his botanical studies; wrote papers on animal fluids, squinting, artesian wells, and the expansion of air; published a book on medical theory; and

was a founding member of the Lunar Society, which advanced pure and applied science, and has been called the most effective provincial group in England's history. McNeil writes to capture Darwin's vitality and breadth of interests and accomplishments; she found that earlier biographies of him were too narrow. The result is a generously detailed account of his life, ideas, and effect on English agriculture, industry, and literature. Extensive bibliography; end notes; index. 308 pp.

McVaugh, Rogers. *Edward Palmer: Plant Explorer of the American West*. Norman: University of Oklahoma Press, 1956.
Biography. As McVaugh points out, the second half of the nineteenth century was the heyday of professional specimen collectors in the New World. Among the greatest of them in North America was Edward Palmer (1831-1911). Between 1853 and 1910 he roved the West, particularly in the Southwest and northern Mexico, assembling the best collections of the times. However, carelessness in documenting specimens by him and his employers damaged his reputation—mistakenly McVaugh believes. This punctiliously documented scholarly review of Palmer's work seeks to correct the mistake. About one third of the text concerns Palmer's life and career. The rest contains a geographical index of his collections, field notes, and last will. Photos; bibliography; footnotes; index. 430 pp.

Manber, David. *The Wizard of Tuskegee*. New York: Crowell-Collier Press, 1967.
Biography for readers 14 years and older. Manber credits George Washington Carver with starting the revolution of farming during the late nineteenth and early twentieth centuries in America by using chemical analysis and scientific growing methods to improve crops and the variety of foods. Carver began his long career in science, teaching, and industry at the Tuskegee Institute, devoting himself to improving the life of fellow African Americans, especially the rural poor, but as Manber shows, his research benefited all Americans. His modesty and perseverance in improving the nation's food supply made him the most famous and honored American

chemist and agricultural experimenter of his time. Photos and drawings; bibliography; index. 168 pp.

Mann, William M. *Ant Hill Odyssey*. Boston, Mass.: Little, Brown, 1948.
Memoirs. In a delightful, often funny narrative William Mann tells of his experiences as an entomologist in the early twentieth century. He remarks briefly about his childhood and then describes his graduate studies and work for the U.S. Bureau of Entomology in detail. He joined expeditions to study insects, especially ants, in Brazil, the Middle East, Haiti, and Polynesia. The book ends when Mann finally lands a permanent job as the ant specialist for the Department of Agriculture. A wonderfully entertaining book for general readers. Photos. 338 pp.

Manning, Kenneth R. *Black Apollo of Science*. New York: Oxford University Press, 1983.
Biography. Ernest Everett Just (1883-1941), a cell biologist specializing in marine embryology, was a complex, brilliant man who struggled throughout his life to find his place as an African American scientist in a white-dominated profession. In Manning's capable hands, Just's story contains much drama and triumph but is finally rather tragic. According to Manning, in addition to the widespread hostility toward African Americans, Just faced constant frustration in securing funding for research, in dealing with the administration at Howard University, where he was professor of zoology, and in finding happiness in his family life. Manning concentrates on academic and funding politics, providing balanced portraits of both scientists who helped Just and those who hindered him. Manning also explains the outlines of Just's research achievements, although readers unfamiliar with the history of biology may have difficulty appreciating the importance of Just's discoveries and ideas. Photos; bibliography; end notes; index. 397 pp.

Meine, Curt. *Aldo Leopold*. Madison: University of Wisconsin Press, 1988.
Biography. Aldo Leopold (1887-1948) characterized an ecolo-

gist as "skillful in seeing facts, ingenious in forming hypotheses, and ruthless in discarding them when they don't fit." Meine takes that standard as his own in this biography of the great conservationist and naturalist. He covers Leopold's family background and education in considerable detail and then focuses on his work for the Forest Service, popular writing, professorship in game management at the University of Wisconsin, and studies of ecology, land use, and government policy. Although it contains the documentation that scholars demand of a formal biography, the book is accessible to general readers. Photos; bibliography; end notes; index. 638 pp.

Metchnikoff, Olga. *Life of Elie Metchnikoff*. Boston, Mass.: Houghton Mifflin, 1921.
Biography. The wife of Ilya Ilich Metchnikov (1845-1916) writes of his single-minded devotion to the improvement of human health. As a staff member of the Institut Pasteur, he studied cell layers and the internal chemical action of cells. His discovery of phagocytes, the blood cells that attack intruders in the bloodstream, brought him the 1908 Nobel Prize for Physiology or Medicine. This biography by his widow explains Metchnikov's biological work and his philosophy of life ("Orthobiosis") for general readers, but it is an idealized portrait of the man and dramatically written. Photo of Metchnikov; bibliography; index. 296 pp.

Mitchell, Peter Chalmers. *Thomas Henry Huxley*. New York: G. P. Putnam's Sons, 1900.
Biography. Written shortly after Thomas Henry Huxley's death in 1895, this book is full of the temper of the times and best regarded as a period piece. Mitchell allots approximately equal space to Huxley's work as a naval surgeon and naturalist, anatomist, taxonomist, champion of Darwinism, philosopher, and educational reformer. He remarks only briefly on Huxley's personal life. Photos; short bibliography. 297 pp.

Moore, James. *The Darwin Legend*. Grand Rapids, Mich.: Baker Books, 1994.
Biography. An example of exhaustive historical detective work, this book debunks the legend that Charles Darwin pro-

fessed his faith in Christianity, rather than science, on his deathbed. Moore not only studies the origin and dissemination of the story, he also comments upon the gullibility of people who accept such stories. Nearly half of the book comprises documents in support of Moore's conclusions. Photos and drawings; bibliography; end notes. 218 pp.

Moore, Ruth. *Charles Darwin*. New York: Alfred A. Knopf, 1958.
Biography. An engagingly written, brief survey of Charles Darwin's life and the theory of evolution. Moore opens with a scene from Darwin's explorations of the Galapagos Archipelago and the focus of his nature studies during the voyage of the *Beagle*. Then she relates his youth, the voyage in greater detail, his marriage and family life, and the development and publication of his ideas. Her final chapter concerns the aftermath of *The Descent of Man* (1871), especially the critical storm that swept over Darwin. A fine first book on Darwin for general readers. End notes; index. 213 pp.

Morwood, William. *Traveler in a Vanished Landscape*. New York: Clarkson N. Potter, 1973.
Biography. This is a pleasantly written biography of the Scots botanist David Douglas and recreates the botanical fervor and temper of the times vividly for general readers. Douglas died in the Hawaiian islands at only 35 years of age, but he had already become famous for his botanizing expeditions in California and, particularly, the Pacific Northwest. He cruised up the Columbia River, sending back dozens of hitherto unrecorded species to the Horticultural Society of London. He is now best known as the namesake of the Douglas fir. This is an informtive introduction not only to a premier nineteenth-century botanist but also to early botany in North America. Drawings; bibliography; index. 244 pp.

Mowat, Farley. *Woman in the Mists*. New York: Warner Books, 1987.
Biography. Dian Fossey (1932-1985) went to Africa when she was thirty-five to live with and study mountain gorillas. An extraordinary story of adventure and frustration, her research occurred despite civil war, primitive living conditions, and

her long battle to save the gorillas from poachers. Poachers murdered her in 1985, but long before that she had established a rapport with the gorillas so that she could learn their behavior in greater depth than any zoologist before. Mowat combines his own research into her life with sections from her journals and publications. A harsh but inspiring story of dedication, which Mowat, a best-selling nature writer, tells dramatically and affectingly. Photos; index. 380 pp.

Muir, John. *Nature Writings*. New York: Library of America, 1997.
Autobiography. This volume collects four long works and selected essays by the most famous American naturalist and conservationist, John Muir (1838-1914). Three of the long works—*The Story of My Boyhood and Youth, My First Summer in the Sierra*, and *Stickeen*—are autobiographical. The essays and *The Mountains of California* are primarily natural history and describe such places as Yellowstone and Yosemite. Fundamental texts for environmentalists and beautifully written. A detailed chronology of Muir's life concludes the text. Drawings and diagrams; end notes; index. 888 pp.

Murphy, Lawrence R., and Dan Collins. *The World of John Muir*. Stockton, Calif.: Holt-Atherton Pacific Center for Western Studies, 1981.
Biography-history. The editors of *The Pacific Historian* offer nine essays by historians about John Muir's life and ideas. Among the topics are his wilderness aesthetics, the formation of the Sierra Club, his family life, and his trip to Alaska. Photos and drawings; footnotes. 91 pp.

Nardo, Don. *Charles Darwin*. New York: Chelsea House, 1993.
Biography for readers 12 years and older. Nardo opens with vivid descriptions of Charles Darwin's adventures as a naturalist during the voyage of the *Beagle* and the discoveries, especially on the Galapagos Islands, that inspired his theory of evolution. The travel adventure turns into an adventure of ideas in Nardo's treatment as he recounts Darwin's life and the struggle to establish natural selection as a reputable scientific principle. An exciting book for young readers interested in science. Photos and drawings; index. 111 pp.

Nisbett, Alec. *Konrad Lorenz*. New York: Harcourt Brace Jovanovich, 1976.
Biography. Konrad Lorenz (1903-1989) studied the behavior of birds and shared the 1973 Nobel Prize for Physiology or Medicine for his work. Late in his career the Austrian ethologist gained wide popular notice by applying his findings in animal behavior to human behavior, especially his assertion that innate aggression can be controlled if understood properly. Drawing in part from conversations with Lorenz, Nisbett relates the major events in his life and career, but the focus of the book rests upon the ideas, and controversy, about behavior that constellated around Lorenz. The discussion of ideas is often sophisticated; nonetheless, the book is aimed at a general audience. Photos; bibliography; index. 240 pp.

O'Brian, Patrick. *Joseph Banks, a Life*. London: C. Harvill, 1987.
Biography. Joseph Banks (1743-1820) accompanied Captain James Cook on his first voyage of exploration, assembled great collections of plant life from around the world, helped design King George III's experimental gardens at Kew, and served as president of the Royal Society for decades. O'Brian describes Banks' explorations in bountiful detail, quoting at length from his journals and letters to bring out the full savor of this extraordinary botanist's intellect and undaunted energy. O'Brian also explains the political and social milieu in which Banks rose to become the most powerful scientist in England in his times. Best of all, at least for botanists, O'Brian discourses on the passion moving natural philosophers of the eighteenth century to comb the world for specimens in order to preserve, display, categorize, and, most of all, wonder at the diversity of nature. Portraits; bibliography; end notes; index. 328 pp.

Orel, Vítezslav. *Gregor Mendel, the First Geneticist*. Oxford, England: Oxford University Press, 1996.
Scientific biography. Orel examines the social and cultural milieu in which Gregor Mendel (1822-1884) performed the experiments that helped him formulate laws of heredity. Orel explains how much was understood about heredity before Mendel and then reviews his childhood and education.

Thereafter, the book delves into the mechanics of Mendel's experiments with peas, the political situation at Mendel's monastery in Brno (now in the Czech Republic) and its effects on him, and the reception of his theory of heredity. Readers familiar with basic genetics and the history of biology will profit most from the book. Photos and diagrams; bibliography; index. 363 pp.

Orel, Vítezslav. *Mendel*. Oxford, England: Oxford University Press, 1984.
Biography. A much more concise, less technically sophisticated version of *Gregor Mendel* (1996, see the entry immediately above). The plan of the book is the same, except that Orel does not trace the acceptance of Mendel's theory of heredity. The text explains the famous pea experiments in detail, but it is still well suited to general readers interested in science history. Bibliography; index. 111 pp.

Outram, Dorinda. *Georges Cuvier: Vocation, Science and Authority in Post-Revolutionary France*. Manchester, England: Manchester University Press, 1884.
Biography. Outram scrutinizes the life of Georges Cuvier (1769-1832) in order to assess how the worlds of science and politics interacted in early nineteenth-century France and how scientists became part of the new government class of the period. The author's text accordingly focuses on Cuvier's scientific and public careers after a review of his early life. Cuvier's work in biology and geology receive attention especially in connection to the controversies and conflicts he created with other scientists, principally Lamarck. The closely argued themes are intended for other historians of science. Large bibliography; end notes; index. 299 pp.

Pauly, Philip J. *Controlling Life: Jacques Loeb and the Engineering Ideal in Biology*. New York: Oxford University Press, 1987.
Scientific biology. Pauly addresses the central motivating principle in the career of Jacques Loeb (1859-1924), a star of biological research in late nineteenth- and early twentieth-century America. Loeb believed that physiological research would soon allow scientists to control and shape organisms.

He viewed biologists as engineers rather than experimenters or theorists. His view did produce some concrete results: In 1899 he produced the first artificial parthenogenesis, a feat that earned him popular fame and opprobrium. Pauly skims over the details of Loeb's personal life in the introduction, for the most part, and focuses on the European origins of Loeb's views and techniques and his influence on later biologists, which was extensive. Among his students was the Nobel laureate geneticist Hermann Muller, for instance. The text is for readers familiar with American science history and basic biology. Photos and drawings; end notes; index. 252 pp.

Pearson, Hesketh. *Doctor Darwin*. New York: Walker, 1930.
Biography. Pearson writes of Erasmus Darwin, Charles Darwin's grandfather. The elder Darwin, the author argues, invented or foresaw the invention of nearly everything in the modern world, including evolution. A descendent of Darwin, the author claims to draw upon unpublished material. The book is for general readers, and there is never any doubt of the author's partisanship. Much of the book discusses Darwin's many friends in the Lunar Society, where members discussed pure and applied science, but the narrative also addresses Darwin's botanical, biological, and medical ideas. A portrait of Darwin; bibliography; index. 244 pp.

Piternick, Leonie K., ed. *Richard Goldschmidt, Controversial Geneticist and Creative Biologist*. Basel, Switzerland: Birkhäuser, 1980.
Biography-science review. Seven of the essays in this volume concern the genetic and zoological research of Richard B. Goldschmidt. There are three reprints of famous articles by him and a reprint of a substantial biographial sketch of him by a colleague at the University of California Department of Zoology, Curt Stern. Most of the text requires extensive knowledge of biology and genetics to follow. Photos; end notes for each article. 154 pp.

Plate, Robert. *Alexander Wilson, Wanderer in the Wilderness*. New York: David McKay, 1966.
Biography. For general readers, Plate emphasizes the mystery

and romantic adventure in the life of Alexander Wilson, often called the father of American ornithology. He was also a poet and painter, using his visual powers beautifully in his book *American Ornithology*. Complaining that Wilson's reputation is unjustly obscured by that of John James Audubon, Plate sets out to revive it, but he seems to have trouble deciding why Wilson wandered in the wilderness studying and painting birds. Plate finally decides that Wilson saw it as a patriotic gesture to bring harmony to his adopted country. Drawings; bibliography; index. 216 pp.

Pratt, Paula Bryant. *Jane Goodall.*. San Diego, Calif.: Lucent Books, 1997.
Biography for readers 12 years and older. Jane Goodall (b. 1934) won the hearts of nature lovers with her books—and in the documentaries made about her—describing her work with chimpanzees. The story of her long, close observations of primates in Africa is dramatic, ideal for young readers, and well presented by Pratt, who stresses that Goodall's method of study by living close to her subjects and familiarizing them with her was an innovation. The book relates Goodall's youth, education, and career in Africa, and sidebar quotations from her writings and testimonials from others accompany the text. Photos; bibliography; end notes; index. 110 pp.

Pringle, Laurence. *Wolfman: Exploring the World of Wolves*. New York: Charles Scribner's Sons, 1983.
Biography for readers 12 years and older. Pringle follows the career and field research of Dave Mech (b. 1937). Beginning in 1958 Mech studied wolf packs in national parks. His work helped to dispel some of the misconceptions about the hunting behavior of wolves, supporting the theory that wolves help cull herds of sick and disabled animals. Pringle offers young readers a model researcher in Mech, courageous and resourceful. An entertaining and enlightening book. Photos; bibliography; index. 71 pp.

Provine, William B. *Sewall Wright and Evolutionary Biology*. Chicago, Ill.: University of Chicago Press, 1986.
Scientific biography. Provine's scholarly consideration of

Sewall Wright (1889-1988) not only reveals the ideas of a pioneer in genetics and evolutionary biology but it is practically a history of these specialties, because Provine is careful to place Wright's research in its context. Wright's career lasted three-quarters of a century. Including some of the mathematics of genetics as well as technical explanations, this book is for readers familiar with the basics of biology and evolution theory. Photos; bibliography; index. 545 pp.

Raven, Charles E. *John Ray, Naturalist: His Life and Works*. Cambridge, England: Cambridge University Press, 1986.
Biography. First published in 1942, this erudite book recounts the career of John Ray (1627-1705). Raven claims that Ray, an early botanist and zoologist, laid the foundation for the modern outlook on nature as valuable in its own right and a proper object of inquiry. Ray is in fact an admirable figure. A blacksmith's son, he acquired an education but left Trinity College rather than sign an oath of allegiance to Oliver Cromwell's Covenant. He then toured Britain and Europe studying plants, birds, and animals. He published fifteen books on his findings and the relation between science and religion. Ray's *History of Plants* has been called one of the most important botanical works by an English author. Portrait of Ray; footnotes; indexes. 506 pp.

Roberts, Jack. *Dian Fossey*. San Diego, Calif.: Lucent Books, 1995.
Biography for readers 12 years and older. A premier example of mid-career change, Dian Fossey left her job as a physical therapist to study mountain gorillas in Africa. As Roberts makes clear, she not only brought much new knowledge and discredited myths about them; she also made the world aware of the dangers to them from poachers. Her work, writings, and documentaries, in fact, may have saved the mountain gorillas but led to her murder in 1985. A poignant story, and Roberts presents it evocatively for young readers. Sidebars with quotations from Fossey's writings accompany the narrative. Photos; bibliography; end notes; index. 111 pp.

Rodgers, Andrew Denny III. *John Torrey*. New York: Hafner Publishing, 1965.

Biography. A facsimile of the 1942 edition, this book describes the career of John Torrey (1796-1873). A chemistry professor at Columbia University and then Princeton University, Torrey founded two famous herbaria and was a pioneer in taxonomic botany in the United States. Often linked to Asa Gray for their botanical explorations of the West, Torrey was an estimable scientist in his own right, Rodgers shows. In great detail, he covers Torrey's lineage, youth, education, system of plant classification, herbaria, botanical club, views of Gray and Louis Agassiz, and explorations in New York, Florida, and the far West. Photo of Torrey; bibliography; index. 352 pp.

Rodgers, Andrew Denny III. *"Noble Fellow": William Starling Sullivant*. New York: Hafner Publishing, 1968.
Biography. First published in 1940, this study of William Starling Sullivant (1803-1873) explains why he is considered the "father of bryology"—that is, the study of mosses. A long introductory chapter about his family and their moves in the South prefaces a close account of Sullivant's education at the University of Ohio, presidency of a Columbus bank, botanical interests, correspondence with John Torrey and Asa Gray, collecting trips for moss specimens in the Allegheny Mountains, construction of a herbarium, and publications. The confusingly designed documentation system of this book makes it an awkward resource for scholars and may annoy lay readers. Photos; bibliography; index. 361 pp.

Rothra, Elizabeth Ogren. *Florida's Pioneer Naturalist: The Life of Charles Torrey Simpson*. Gainesville: University Press of Florida, 1995.
Biography. A marine biologist, Charles Torrey Simpson (1846-1932) specialized in mollusks, naiads in particular. When he retired from the Smithsonian's National Museum, he moved to Florida and began an long exploration of the wilds. He studied the exotic plant life, writing both scholarly and popular accounts. Rothra covers his early life fairly quickly—from Simpson's birth in Illinois through his tenure at the Smithsonian—in order to concentrate on his years collecting specimens in Florida and the keys. Photos; bibliography; end notes; index. 232 pp.

Rourke, Constance. *Audubon*. New York: Harcourt, Brace, 1936.
Biography. This awkwardly written but pleasingly illustrated book refers to John James Audubon's personal life in roundabout, dainty terms, probably in order not to offend the sensibility of readers of the 1930's. Rourke's descriptions of natural settings and Audubon's art are occasionally vivid despite the general passive voice texture of the prose, as the book concentrates on the explorations of America in search of birds. Drawings and artwork; bibliographical essay; index. 342 pp.

Rupke, Nicolaas A. *Richard Owen: Victorian Naturalist*. New Haven, Conn.: Yale University Press, 1994.
Biography. Richard Owen (1804-1892) was one of the most distinguished English scientists of the nineteenth century. He published books on zoology and paleontology, directed museums, became a member of the Royal Society, was knighted, knew most of the leading naturalists of his day, and held several professorships, including the Fullerian professorship at the Royal Institution. Rupke devotes chapters to specific aspects of Owen's long , varied career, such as his affiliation with museums, his scientific accomplishments, and his debates with Thomas Henry Huxley about humankind's place in nature. Accordingly, he does not present the facts of Owen's life chronologically; however, a chronological list of important events prefaces the text. Best suited for historians of science. Drawings and diagrams; bibliography; end notes; index. 462 pp.

Scheffer, Victor B. *Adventures of a Zoologist*. New York: Charles Scribner's Sons, 1980.
Autobiography. Marine biologist, diplomat, popular science writer, and photographer, Victor Scheffer led a varied, exciting life. His studies of seals in the Pribilof Islands were landmarks in the population biology of marine mammals, and in *The Year of the Whale* (1969), a best-seller, he helped make the public aware of the dangers whales face from people. But the most remarkable part of his long career, as he relates it in this book, is his change of heart as a zoologist. He began as a wildlife manager who preserved animals for efficient harvest and use by people; gradually, exploitation and harvesting

repelled him and he became a dedicated preservationist and ecologist. This autobiography, while thoughtful, is unfortunately desultory and sometimes simply a series of anecdotes; nevertheless, it illustrates a change in the American view of nature. Photos; index. 204 pp.

Schierbeek, A. *Jan Swammerdam*. Amsterdam, The Netherlands: Swets and Zeitlinger, 1967.
Biography. Jan Swammerdam (1606-1680) was a naturalist and anatomist who dissected, described, and drew insects of northern Europe, the work for which he is best known. However, he also published analyses of human and animal anatomy. Schierbeek opens this study with a biographical sketch of Swammerdam and then devotes nine chapters to his work in physiology, anatomy, entomology, and botany. The book is an overview of Swammerdam for use by other scholars. Drawings; bibliography; footnotes; index. 202 pp.

Schierbeek, A., with Maria Rooseboom. *Measuring the Invisible World*. London: Abelard-Schuman, 1959.
Biography. Only the first chapter—about one-tenth of this book—discusses the life of Anton van Leeuwenhoek, a spare sketch of the inventor of the microscope. The rest of the book reviews in detail the development of Leeuwenhoek's thought and method of solving problems. Specifically, the text concerns his microscopes, microbiology, theory of generation, and studies of insects, plants, animals, and chemistry. Photos and drawings; bibliography; index. 223 pp.

Sears, Paul B. *Charles Darwin: The Naturalist as a Cultural Force*. New York: Charles Scribner's Sons, 1950.
Biography. To Sears, Charles Darwin did much more than advance the theory of evolution and unveil first principles. He showed that an individual, trusting his own eyes and mind, can by "open trial and honest evidence" arrive at the truth. That the truth pried apart the assumptions upon which Victorian society was based made Darwin's work heroic. Sears argues, however, that before Darwin could free the minds of others from error, he had to free his own, and in a somewhat labored style Sears offers to show readers how he

managed it. Chapters address Darwin's youth, his voyage on the *Beagle* around the world, publication of *The Origin of Species* (1859), and the reception of natural selection during Darwin's life and afterward. Bibliography; index. 124 pp.

Skutch, Alexander F. *The Imperative Call.* Gainesville: University Presses of Florida, 1979.
Autobiography-popular science. Alexander F. Skutch (b. 1904) tells how in his personal and professional life he followed the voice of nature. Specifically, he followed the sounds of birds, his specialty. Although he recalls his youth and education, explaining how he became a birdwatcher, most of the text involves the birds themselves. His experiences in the United States, Jamaica, and Central America (particularly Guatemala) are occasions for discussions of the habitats, behavior, and appearance of such species as jays, motmots, anis, and kingfishers. Photos; index. 331 pp.

Soyfer, Valery N. *Lysenko and the Tragedy of Soviet Science.* New Brunswick, N.J.: Rutgers University Press, 1994.
Biography-history. Soyfer, a biophysicist trained in the Soviet Union, tells a chilling story of science perverted by politics. Trofim Lysenko (1898-1976) was a "semiliterate, self-taught amateur" who rose to prominence based upon fraudulent reports of experiments with wheat. Ambitious far beyond his scientific training, he proposed means for improving Soviet agriculture in the 1930's, which Joseph Stalin eagerly seized upon as an escape from chronically poor harvests. Lysenko rose to high political position and used it to squelch his scientific enemies, particularly geneticists. The result, says Soyfer, crippled biological science in Russia for decades, and the phenomenon of "Lysenkoism" revealed the weakness of Soviet science in general, where conformance to doctrine and pleasing party leaders were more important than verifiable knowledge. Soyfer collected documents for more than two decades in preparation for this book and during his student days interviewed Lysenko himself. Soyfer stops short of depicting Lysenko as scientific evil incarnate; still, he illustrates the dangers of a Machiavellian science administrator to science as a whole and society. Photos; end notes; indexes. 379 pp.

Steinbeck, John. "About Ed Ricketts." In *Log from the Sea of Cortez*. New York: Viking, 1951; Penguin Books, 1976: vii-lxiv.
Biography. Steinbeck writes affectionately about his friend and boss at a marine research station in Monterey, California. Edward Ricketts was the model for Doc in Steinbeck's novel *Cannery Row*, and this portrait shows him to be even more eccentric than the fictional character. Steinbeck spends little space on Ricketts' scientific work, although he was a prolific collector of littoral sea life. Instead, Steinbeck describes a maverick intellectual working in a small town during the Depression, an American type. The long narrative following the portrait does provide glimpses into Ricketts' pursuit of marine biology as he and Steinbeck tour the Gulf of California to collect specimens. Index.

Sterling, Keir B. *Last of the Naturalists*. New York: Arno Press, 1977.
Biography. Clinton Hart Merriam (1855-1942) was the first director of the U. S. Biological Survey, which later evolved into the Department of Agriculture. A physician by training, he contributed to the fields of mammalogy, ornithology, and biogeography. Sterling gives some attention to Merriam's personality and background but devotes the text to his work as a naturalist and government scientist. To Sterling, Merriam represents a transition from the explorers and collectors who predominated in biology through the first half of the nineteenth century and the specialists of the twentieth century. The audience is primarily science historians. Photos; bibliography; end notes; index. 472 pp.

Streshinsky, Shirley. *Audubon: Life and Art in the American Wilderness*. New York: Villard Books, 1993.
Biography. A well-written account of Audubon's turbulent youth, maturity in the United States, innovations in taxidermy and wildlife art, and contribution as a naturalist and ornithologist. Almost a synonym for American wildlife studies, Audubon published one of the nation's most influential books, *The Birds of America*, and Streshinsky recounts the genesis of the project in detail. She also explains the difficulties in reconstructing Audubon's adventurous life because his family

bowdlerized some of his journals and letters. Photos and artwork; bibliography; end notes; index. 407 pp.

Teale, Edwin Way, ed. *The Wilderness World of John Muir*. Boston, Mass.: Houghton Mifflin, 1954.
Biography. In his introduction, Teale discusses Muir's interest in geology, discovery of living glaciers in the Sierra Nevada, studies of trees, and theory of land formation, but he views Muir as a larger-than-life visionary and philosopher more than as a scientist. The rest of the book excerpts from Muir's writings; in prefaces to each section Teale provides biographical and interpretive comments. Drawings; index. 332 pp.

Turner, Frederick. *Rediscovering America*. New York: Viking, 1985.
Biography. Turner opens this entertainingly written biography describing the ten-year-old John Muir's excited reaction to news of a gold strike in California and his fantasies about the vast American wilderness. These two facets of the nation cunningly introduce basic themes to Muir's story: the tension between exploitation of resources and preservation of natural beauty. In the first three parts of the book, Turner follows Muir from his birthplace, Scotland, to America and his wanderings, geologizing, and writing in the Sierra Nevada. In the final part, he discusses Muir's legacy to American forests and national parks. Photos; end notes; index. 417 pp.

Twitty, Victor Chandler. *Of Scientists and Salamanders*. San Francisco, Calif.: W. H. Freeman, 1966.
Scientific biography. Trained as an embryologist at Yale University, Twitty wound up studying salamanders during his long career at Stanford University. He touches upon his education and career, but the narrative concentrates on his research in the anatomy, distribution, and behavior of the amphibians, which Twitty explains at a level best understood by a fellow biologist. Photos and drawings; bibliography; index. 178 pp.

Uggla, Arvid Hj. *Linnæus*. Stockholm: Swedish Institute, 1957.
Biography. This pamphlet celebrates the 250th anniversary of

Carl Linnaeus' birth. The spare text reviews the principal events of Linnaeus' life and touches very briefly upon his system of botanical nomenclature. Photos and drawings. 34 pp.

Vermeij, Geerat. *Privileged Hands: A Scientific Life.* New York: W.H. Freeman, 1997.
Autobiography. A paleobiologist and evolutionary biologist, University of California professor Geerat Vermeij (b. 1946) edits prestigious professional journals, has won grants from the MacArthur and Guggenheim foundations, and authored several books on evolution and biology. His research has focused on ancient and contemporary mollusks. A stellar modern science career, but what makes the story particularly inspiring is that Vermeij is blind. His well-written narrative reveals how he learned to live with his blindness, incurred at age four, and developed his interest in science. His keen sense of touch turns out to give him an advantage over researchers who rely on their eyes alone for analyzing specimens. An absorbing book for all readers interested in biology or blindness. Photos; glossary; index. 297 pp.

Wadsworth, Ginger. *Rachel Carson, Voice for the Earth.* Minneapolis, Minn.: Lerner Publications, 1992.
Biography for readers 12 years and older. A brilliant popularizer and synthesizer, Rachel Carson (1907-1964) influenced America's environmental movement profoundly with *The Sea Around Us* (1951) and, especially, *Silent Spring* (1962), her indictment of pollution from commercial fertilizers. She was trained as a marine biologist and worked for the U.S. Fish and Wildlife Service, as Wadsworth recounts in this inspirational portrait of a tireless worker. Photos and drawings; brief bibliography; end notes; index. 128 pp.

Waters, Kenneth C., and Albert Van Helden, eds. *Julian Huxley: Biologist and Statesman of Science.* Houston, Tex.: Rice University Press, 1992.
Biography-history. After a general introduction sketching the life and career of Julian Huxley, 18 essays analyze his place in the English intelligentsia, contributions to biology, and work in science politics and popularization. The volume publishes

papers delivered at a conference sponsored by Rice University, where Huxley once taught. The authors are historians and scientists; they write primarily for other scholars. Photos and diagrams; substantial bibliography; end notes; index. 344 pp.

Weiss, Harry B., and Grace M. Ziegler. *Thomas Say, Early American Naturalist*. Springfield, Ill.: Charles C Thomas, 1931.
Biography. The authors call Thomas Say (1787-1834) the father of American zoology. He was an early collector of specimens, especially shells and insects, and his collections were widely studied; he also wrote on topics in paleontology. The authors devote a long chapter to Say's family and another to Say's early life; further, shorter chapters concern his travels, life in the New Harmony experimental commune, friends, scientific achievements (sketchily described), and extensive collections. Drawings; bibliography; index. 260 pp.

Welker, Robert Henry. *Natural Man: The Life of William Beebe*. Bloomington: Indiana University Press, 1975.
Biography. A foremost naturalist of the twentieth century, William Beebe (1877-1962) specialized in tropical fauna and birds. Welker describes Beebe's travels to Borneo, the Galapagos Islands, Guinea, and the Caribbean and collections of bird specimens. Beebe also went underwater—in hard-hat suits and a bathysphere—to study fish. Welker explains in detail the background and methods of Beebe's scientific work and also discusses his many popular books about nature. Photos and drawings; bibliography; end notes; index. 224 pp.

Wichler, Gerhard. *Charles Darwin: The Founder of the Theory of Evolution and Natural Selection*. Oxford, England: Pergamon Press, 1961.
Biography-history. For students of Charles Darwin's ideas, Wichler summarizes their historical and intellectual background. The book's first part concerns various approaches to evolution before Darwin. The second part explains Darwin's theories of selection and descent and how he arrived at them. The final part takes up Darwin's personal life briefly, considers the influence of his friends and collaboration with Joseph

Hooker, and reviews Darwin's biological and psychological studies. Photos and drawings; short bibliography; footnotes; indexes. 228 pp.

Wilkins, Thurman. *John Muir: Apostle of Nature.* Norman: University of Oklahoma Press, 1995.
Biography. Wilkins shows in this concise, perceptive account that John Muir formed his philosophy of nature early. Among its tenets were that all life is sacred and that wild nature should be exalted over civilization. Such beliefs were bound to land him in conflict, and they did as he fought to preserve wild areas in California, a fight he sometimes lost. From his boyhood in Scotland to his death in California, Muir was something of a paradoxical character, according to Wilkins, who emphasizes the practical side of Muir rather than his romantic, mystical tendencies. Thus, the reader sees Muir primarily as the figure who started conservationism, although Muir's biology and geology also receive attention. Photos and drawings; bibliographical essay; index. 302 pp.

Wilson, Edward O. *Naturalist.* Washington, D.C.: Island Press, 1994.
Autobiography. "Earth, in the dazzling variety of its life, is still a little-known planet," writes E. O. Wilson toward the end of this elegantly composed book. In it he recounts his attempts as an entomologist and evolutionary biologist to increase the human awareness of fellow species in the environment. Wilson is a synthesist of ideas as well a collector of data. In addition to his many discoveries about ants, he helped found sociobiology and the biodiversity movement. He explains how these ideas developed and the strident controversy over sociobiology. The book thus reveals both a strikingly dedicated and penetrating thinker and the social texture of academic science during his career at Harvard University, which began in the 1950's. He also tells how his boyhood fascination with insects led to that career, among the most honored in American science. Photos and drawings; index. 380 pp.

Wolfe, Linnie Marsh. *Son of the Wilderness: The Life of John Muir.* New York: Alfred A. Knopf, 1945; Madison: University of

Wisconsin Press, 1978.
Biography. This gently written book takes readers through John Muir's life decade by decade. Having carefully traced down many rumors and libels about Muir, Wolfe refutes the canards and clarifies the legends. But she warns that readers who know Muir only as a naturalist and mystic miss much of interest. He was no plaster saint, she insists, but a man of extraordinary energy, practicality, humor, and, in a fight, fierceness. The book concludes with Muir's last great fight—over the Hetch Hetchy reservoir in the Sierra Nevada, which the author claims shortened Muir's life and reduced his literary output. Photos and drawings; bibliography; end notes; index. 380 pp.

Chapter 6

Mathematics

COLLECTIONS

Albers, Donald J., and Gerald L. Alexanderson, eds. *Mathematical People*. Boston, Mass.: Birkhäuser, 1985.
Profiles and interviews with 25 contemporary mathematicians, including Paul Erdös, Benoit Mandelbrot, and Stanislaw Ulam. The editors' want to show that mathematicians have as wide a variety of personalities as other professional people. The interviews are in question-and-answer format, and each comes with a short biographical sketch. A historical essay that links mathematics and culture prefaces the profiles. Photos, drawings, and diagrams; index. 372 pp.

Albers, Donald J., Gerald L. Alexanderson, and Constance Reid. *More Mathematical People: Contemporary Conversations*. San Diego, Calif.: Academic Press, 1994.
This volume offers brief biographical information about and extended interviews with 18 influential mathematicians, most from academia but some from industry as well. The interviews are historically valuable because they contain firsthand information on the development of twentieth-century mathematics and computer science. They also reveal the diverse personalities of the interviewees and show students how to become a mathematician and the rewards of the career. Photos and drawings; index. 375 pp.

Bell, Eric Temple. *Men of Mathematics*. New York: Simon and Schuster, 1937.

For readers who can follow demonstrations of geometry and algebra, Bell offers biographical sketches of leading mathematicians and summaries of their mathematical innovations. He begins with the Hellenic thinkers Zeno (fifth century B.C.E.), Eudoxus (408-355 B.C.E.), and Archimedes (287-212 B.C.E.) and concludes with Georg Cantor (1845-1918), featuring 29 individual mathematicians and one mathematical family (the eight Bernoullis). With these profiles, Bell outlines the history of Western mathematics. Photos, drawings, and diagrams; index. 592 pp.

The Biographical Dictionary of Scientists: Mathematicians. New York: Peter Bedrick Books, 1986.
A survey of the history of mathematics prefaces short articles on 178 mathematicians and theoretical physicists, as early as Thales of Miletus (ca. 624-547 B.C.E.). Averaging about 500 words in length, entries supply basic biographical information and summaries of scientists' mathematical research for a general audience. A helpful resource because of the extent of coverage, brevity of the articles, and well-written glossary. Index. 204 pp.

Fang, J. *Mathematicians, from Antiquity to Today.* Hauppauge, N.Y.: Paideia Press, 1972.
Following a long, segmented introduction on the history and philosophy of mathematics, Fang provides biographical summaries or short essays of up to a thousand words on more than 500 mathematicians. All entries give birth and death dates, country of origin, and some indication of the person's contribution to mathematics. 341 pp.

Grinstein, Louise S., and Paul J. Campbell, eds. *Women of Mathematics.* New York: Greenwood Press, 1987.
This collection of essays provides biographical sketches of 43 women; each sketch is accompanied by a summary of the woman's work in mathematics, a bibliography of her publications, and an annotated bibliography of publications about her. The sophistication of the mathematics explained ranges widely, but generally readers with basic algebra and geometry will understand the text. This book includes more mathemati-

cians and lists more information resources than other reference works about women mathematicians for general readers. Indexes. 292 pp.

Halmos, Paul R. *I Have a Photographic Memory*. Providence, R.I.: American Mathematical Society, 1987.
Halmos presents more than 600 snapshots of mathematicians, taken between 1941 and 1986. Each photograph comes with a blurb containing Halmos' personal remarks about the mathematician. Index. 313 pp.

Henderson, Harry. *Modern Mathematicians*. New York: Facts on File, 1996.
Suited to readers 12 years and older. Emphasizing that mathematics is the study of patterns, Henderson explains the work of 13 mathematicians, as well as sketching their lives, from the nineteenth and twentieth centuries: Charles Babbage (1792-1871), Ada Lovelace (1815-1852), George Boole (1815-1864), Georg Cantor (1845-1918), Sophia Kovalevsky (1850-1891), Emmy Noether (1882-1935), Srinivasa Ramanujan (1887-1920), Stanislaw Ulam (1909-1984), Shiing-Shen Chern (b. 1911), Alan Turing (1912-1954), Julia Bowman Robinson (1919-1985), Benoit Mandelbrot (b. 1924), and John H. Conway (b. 1937). While the selection may omit some celebrated mathematicians, it introduces young readers to topics that may appeal to them, including the mathematics of computers and games. Photos; bibliography for each chapter; index. 139 pp.

Henrion, Claudia. *Women in Mathematics: The Addition of Difference*. Bloomingdale: University of Indiana Press, 1977.
Henrion matches 11 biographical sketches of contemporary women mathematicians with five essays that discuss the myths, politics, and abilities of women in the discipline. Those profiled include Joan Birman, Lenore Blum, Fan Chung, Marcia Groszek, Fern Hunt, Linda Keen, Vivienne Malone-Mayes, Marian Pour-El, Judy Roitman, Mary Ellen Rudin, and Karen Uhlenbeck. Initially, Henrion writes in her introduction, she wanted a book that would encourage more women to become mathematicians. She found, however, that she had to refute the biases discouraging women from entering the

field, and so the book became a study of misconceptions as well. Photos; bibliography; end notes; index. 293 pp.

Hollingdale, Stuart. *Makers of Mathematics*. London: Penguin Books, 1991.
The author supplies a history of mathematics by briefly reviewing biographical information on its most famous practitioners and then analyzing their greatest contributions for readers with an interest in the subject but no advanced training. The book's chapters concern the beginnings of mathematics; early Greek mathematics; Euclid and Apollonius; Archimedes and the later Hellenistic period; medieval Eastern and Western studies; Renaissance mathematics; René Descartes, Pierre de Fermat, and Blaise Pascal; Isaac Newton; Newton's *Principia mathematica*; Newton's followers, such as Isaac Barrow and Edmond Halley; Gottfried Wilhelm Leibniz; Leonhard Euler; Jean d'Alembert; Carl Friedrich Gauss; William Hamilton and George Boole; Richard Dedekind and Georg Cantor; and Albert Einstein. A lucid, well-written history of mathematicians through the nineteenth century. Photos, drawings, and diagrams; end notes; index. 439 pp.

Muir, Jane. *Of Men and Numbers*. New York: Dodd, Mead, 1965.
Muir introduces general readers to the greatest mathematicians of all time, in her estimation: Pythagoras (6th century B.C.E.), Euclid (4th century B.C.E.), Archimedes, Girolomo Cardano (1501-1576), René Descartes (1596-1650), Blaise Pascal (1623-1662), Isaac Newton (1642-1727), Leonhard Euler (1707-1783), Carl Friedrich Gauss, Nicholas Lobatchevsky (1793-1856), Évariste Galois (1811-1832), and Georg Cantor. The biographical material is entertainingly written; the summaries of mathematical ideas require knowledge of basic algebra and geometry. Diagrams; bibliography; index. 249 pp.

Osen, Lynn M. *Women in Mathematics*. Cambridge, Mass.: MIT Press, 1974.
Osen relates the lives of Hypatia (370-415), Hroswitha (the "witch of Agnesi," 1718-1799), Emilie du Châtelet (1706-1749), Caroline Herschel (1750-1848), Sophie Germain (1776-1831), Mary Fairfax Somerville (1780-1872), Sophia Kovalevsky

(1850-1891), and Emmy Noether (1882-1935). Osen also describes their work for general readers. The final chapter considers the familial and societal influences that keep women from becoming interested in mathematics and those that encourage them. Photos and drawings; bibliography; index. 185 pp.

Perl, Teri. *Math Equals: Biographies of Women Mathematicians + Related Activities*. Menlo Park, Calif.: Addison-Wesley, 1978.
Suitable for readers 14 years and older. The book presents biographies of nine women mathematicians and related games and puzzles: Hypatia, Emilie du Châtelet, Maria Gaetana Agnesi, Sophie Germain, Mary Fairfax Somerville, Grace Chisholm Young, Ada Byron Lovelace, Sophia Kovalevsky, and Emmy Noether. The last three also appear in *Women and Numbers* (see below). The narrative includes simplified mathematics that illustrate the achievements of the subjects. Photos, diagrams, and drawings. 250 pp.

Perl, Teri. *Women and Numbers*. San Carlos, Calif.: World Wide Publishing/Tetra, 1993.
Perl writes for young women, and the 12 biographical sketches present women who have succeeded in careers as mathematicians and can serve as role models. The mathematics discussed is rudimentary; accompanying puzzles and paint-by-numbers-style activities seek to entertain while teaching mathematical concepts. Photos, drawings, and diagrams. 211 pp.

Pólya, George. *The Pólya Picture Album: Encounters with Mathematicians*. Ed. by G. L. Alexanderson. Boston, Mass.: Birkhäuser, 1987.
Pólya was an avid amateur photographer and collector of occasional photos. Following a biographical sketch of Pólya by Alexanderson, this album shows the mathematicians Pólya knew, and a few physicists, during his long career in European and American universities, and he knew many of the greats: Felix Klein, David Hilbert, G. H. Hardy, John Littlewood, and Paul Erdös were among them. Brief, anecdotal captions accompany some of the photos. Index. 160 pp.

Smith, Steven B. *The Great Mental Calculators.* New York: Columbia University Press, 1983.
Only about one-third of this book contains biographical material. The other two-thirds discusses the psychology and methods of people who have an unusual ability to perform complex calculations without any aids. The biographical section tells the story of 20 calculating prodigies, many of whom were children, from the eighteenth century to the twentieth century. The author's explanations of mathematical techniques require a knowledge of algebra and logarithms. Photos and drawings; bibliography; index. 374 pp.

Stonaker, Frances Benson. *Famous Mathematicians.* Philadelphia, Pa.: J. B. Lippincott, 1966.
Suited to readers 12 years and older. A simply written account of the lives and ideas of Euclid, Archimedes, Aryabhatta, al-Khwarismi, René Descartes, Isaac Newton, Joseph Louis Lagrange, Carl Friedrich Gauss, John von Neumann, and Norbert Wiener. The author explains such mathematical topics as π and the metric system, but the concepts are not developed in detail; even inattentive young readers should not bog down. Index. 188 pp.

Terry, Leon. *The Mathmen.* New York: McGraw-Hill, 1964.
Terry provides a few biographical details and explains the mathematical innovations of nine thinkers in classical antiquity, from the sixth century to second century B.C.E.: Thales, Pythagoras, Plato, Eudoxus, Aristotle, Euclid, Archimedes, Eratosthenes, and Hipparchus. The text carefully analyzes the proofs for geometric theorems at a level suitable for high school students. Diagrams; glossary; index. 222 pp.

Turnbull, Herbert Westren. *The Great Mathematicians.* New York: New York University Press, 1961.
Biography-history. A reprint of the 1921 edition, this book is aimed at general readers who recall their basic algebra and geometry. Turnbull occasionally introduces equations and geometric figures as he discusses the innovations of mathematicians from the Ionian Greek Thales to the Indian genius Srinivasa Ramanujan. Turnbull sketches their personal lives

and focuses on their discoveries and the intellectual milieu. Diagrams. 141 pp.

BIOGRAPHIES AND AUTOBIOGRAPHIES

Akivis, Maks Aizikovich, and Boris Abramovich Rosenfeld. *Élie Cartan*. Providence, R.I.: American Mathematical Society, 1993.
Scientific biography. The French mathematician Élie Cartan (1869-1951) innovated the Lie groups, algebras, differential equations, geometry, and the theory of spaces. The authors devote the first chapter to biographical details. Then they plunge into technical discussions of Cartan's achievements, which only well-trained mathematicians will appreciate fully. An appendix contains Cartan's essay about the French influence on mathematics. Photos; bibliography. 317 pp.

Baum, Joan. *The Calculating Passion of Ada Byron*. Hamden, Conn.: Archon Books, 1986.
Biography. To Baum, Augusta Ada Byron (1815-1852), daughter of the poet Byron and friend of the mathematician and inventor Charles Babbage, was unusually talented and an unusual personality. Like Babbage, who was the first computer designer (the Difference Engine and the Analytic Engine), Byron, later Countess Lovelace, was the first computer programmer because of the mathematical work she did for him. The story of her friendship with the prickly Babbage makes this book worth reading. Baum also considers the forces of early Victorian society and the family influences that both nurtured Byron's scientific interests and frustrated them. For general readers. Photos and drawings; bibliography; end notes; index. 133 pp.

Belhoste, Bruno. *Augustin-Louis Cauchy*. New York: Springer-Verlag, 1991.
Biography. Belhoste claims that, along with Carl Friedrich Gauss, Augustin-Louis Cauchy (1789-1857) was the first truly modern mathematician in that his work, which was vast, has a unifying theme, the search for absolutes. Belhoste places

Cauchy in his times in order to reveal the true fecundity of his genius and his contribution to the development of science. A professor at the École Polytechnique in Paris, Cauchy was often at odds with his times: He was devoutly Catholic and a royalist, and his beliefs led him into exile and prompted him to reject the Enlightenment tenets held by most of his French colleagues. Although Belhoste concentrates mainly on biographical material, when he explains Cauchy's work, he uses advanced mathematics. Photos and diagrams; bibliography; index. 380 pp.

Bellman, Richard. *Eye of the Hurricane.* Singapore: World Scientific Publishing, 1984.
Autobiography. Richard Bellman (b. 1920) writes pugnaciously about his life and career, from his youth in New York City to his work at the Center for Democratic Study in the early 1970's. He also was a professor at Stanford University and the University of Southern California and an analyst at RAND Corporation, where he worked on control theory and the calculus of variations. He says that an autobiography, such as his, is valuable because it lets the reader witness the "inertia, stupidity and prejudice faced by others." Accordingly, he concentrates on academic and corporate politics, and he certainly seizes the opportunity to speak freely and forcefully. Since it contains no formal mathematical demonstrations, the book is for general readers. Index. 344 pp.

Box, Joan Fisher. *R. A. Fisher, the Life of a Scientist.* New York: John Wiley and Sons, 1978.
Biography. A daughter's life of her father, the English mathematician Ronald Aylmer Fisher (1890-1962). A statistician, he pioneered biometrics, the mathematical treatment of genetics, especially the theory of inbreeding, and was a voice in the eugenics movement. The biography elaborates the principles that Fisher developed during his long academic career, during which he held the chair in genetics at University College, London, but the author also provides insight into the private life of her father, drawing not only on her own experience but also the recollections of many of Fisher's colleagues. She sometimes employs mathematics in her expositions, but

college-educated readers can easily understand the text. Photos, drawings, and diagrams; bibliography; end notes; index. 512 pp.

Brewer, James W., and Martha K. Smith, eds. *Emmy Noether: A Tribute to Her Life and Work*. New York: Marcel Dekker, 1981.
Biography and reminiscences. The editors offer both a non-technical biographical sketch of Emmy Noether (1882-1935) and technical discussions of her mathematics. The substantial biographical sketch by Clark Kimberling is followed by a collection of memorials from colleagues. Non-mathematicians can follow both of these sections. However, the final section, concerning her contributions to Galois extensions, the calculus of variations, commutative ring theory, representation theory, and algebraic number theory requires preparation in advanced mathematics. Photos; bibliography; end notes for each essay; index. 180 pp.

Bühler, Walter Kaufmann. *Gauss: A Biographical Study*. New York: Springer-Verlag, 1981.
Scientific biography. Readers will want a close acquaintance with mathematics to appreciate this account of the great German mathematician-astronomer Carl Friedrich Gauss (1777-1855). Frequently quoting from Gauss' writings, the author recounts the political setting of the times, describes Gauss' personal life in detail, and explains his work on arithmetic, number theory, and celestial mechanics. Drawings and diagrams; bibliography; end notes; index. 208 pp.

Dawson, John W., Jr. *Logical Dilemmas: The Life and Work of Kurt Gödel*. Wellesley, Mass.: A. K. Peters, 1997.
Scientific biography. For mathematically sophisticated readers, this book considers the apparent contradictions between the personal life and intellectual achievements of Kurt Gödel (1906-1978). He rocked mathematicians with his incompleteness theorem and impressed Albert Einstein with his analyses of time as a cosmological concept. But Gödel also believed in ghosts and was incapacitated by paranoia late in life. Although widely considered a master modern logician, his ignorance of political conditions in pre-World War II Europe

was as deep as his crush on a cabaret dancer. Accordingly, Dawson finds him to be a "man/child." Photos; index. 368 pp.

DeLacy, Estelle A. *Euclid and Geometry*. New York: Franklin Watts, 1963.
Biography-history. Suited to readers 12 years and older. Considering that only two biographical facts about Euclid (fl. 4th century B.C.E.) are known, a book-length biography of him seems nervy, but DeLacy has produced a lucid, engaging book for young readers interested in math. Most of it concerns cultural and historical background or explains the basics of Euclid's geometry. Occasionally, DeLacy concocts a dramatic scene featuring Euclid to accord with the legends about his temperament and career. While such license does not convey information, it makes for lively reading and sustains an energetic tone. Diagrams; index. 120 pp.

Dick, Auguste. *Emmy Noether, 1882-1935*. Boston, Mass.: Birkhäuser, 1981.
Biography. Dick's sketch of Noether is brief, divided among her childhood and education in Erlangen, Germany, her struggle to teach at Götttingen, and her teaching at Bryn Mawr and Princeton after she fled Nazi persecution. Most of the book consists of reprinted obituaries by Noether's many admirers. Her development of abstract axiomatic algebra receives superficial description for a general audience. Photos; bibliography; index. 193 pp.

Dijksterhuis, E. J. *Archimedes*. Copenhagen, Denmark: Ejnar Munksgaard, 1956.
Biography. Most of this volume contains the works of Archimedes (287-212 B.C.E.). Introductory essays outline his life, explain his achievements in physics and mathematics, and discuss the extant manuscripts of his works. General readers, especially those who enjoy geometry, can understand the text, although it is a scholarly edition. Diagrams; index. 422 pp.

Dunnington, G. Waldo. *Carl Friedrich Gauss: Titan of Science*. New York: Exposition Press, 1955.

Biography. Dunnington concentrates on the career of Carl Friedrich Gauss, outlining the discoveries of this brilliant mathematician without elaborating them in mathematical detail. He also describes Gauss' extensive contributions in astronomy, crystallography, optics, electromagnetism, and geodesy. Pleasant reading for a general audience. An appendix supplies a genealogy. Photos and drawings; bibliography; index. 497 pp.

Glimm, James, John Impagliazzo, and Isadore Singer, eds. *The Legacy of John von Neumann*. Providence, R.I.: American Mathematical Society, 1990.
Scientific biography. Most of this book contains professional essays on the mathematics of John von Neumann (1903-1957), such as the mathematical foundation of computers, game theory, and rings of operators. The first four essays, however, present brief reminiscences of him, including one by his daughter, Marina von Neumann Whitman, and an explanation of his philosophy. Photos and drawings; references for each essay. 324 pp.

Hall, Tord. *Carl Friedrich Gauss*. Cambridge, Mass.: MIT Press, 1970.
Biography. Among the all-time greatest mathematicians, Carl Friedrich Gauss contributed to number theory, algebra, function theory, Euclidean and non-Euclidean geometry, probability theory, astronomy, mechanics, geodesy, hydrostatics, electrostatics, magnetism, and optics. Hall's biography is brief, and as do most biographers of mathematicians, he frequently declines to present a full explanation of mathematical ideas because of space limitations and the specialized knowledge required of readers. Still, he manages to explain a good deal of Gauss's major contributions during a sketchy account of his life. Diagrams; bibliography; index. 176 pp.

Halmos, Paul R. *I Want to Be a Mathematician*. New York: Springer-Verlag, 1985.
Autobiography. In a chatty style, Paul Halmos tells of his career, which included positions at the Institute for Advanced Study and the universities of Chicago, Illinois, Michigan, and

Indiana. His mathematical specialties included measure theory, Boolean algebra, and Hilbert spaces: these topics Halmos explains in technical detail, and readers will often need a knowledge of advanced mathematics to follow the text. The book also discusses academic politics and muses on what it means to be a mathematician. Photos; index. 421 pp.

Hankins, Thomas L. *Jean d'Alembert*. Oxford, England: Clarendon Press, 1970.
Biography-history. Hankins' aim in this scholarly book is to parse the relation between science and philosophy during the Enlightenment by concentrating on a prominent spokesman of the period, Jean d'Alembert (1717-1783). Accordingly, the book devotes much space to the philosophical developments of the times, especially in relation to Isaac Newton's descriptions of energy and motion, although it also contains a great deal of information about d'Alembert and his mathematics. He attempted to revise Newton's laws of motion using different assumptions and studied the behavior of mechanical systems subject to restraints, rather than mass particles as Newton had done. Hankins employs sophisticated mathematics in the text. Diagrams; bibliography; index. 260 pp.

Hardy, Godfrey Harold. *A Mathematician's Apology*. Cambridge, England: University of Cambridge Press, 1940; Canto, 1992.
Autobiography. Only the last chapter of this brief book is autobiographical, and even that chapter regards only the matters that led G. H. Hardy (1877-1947) to become one of England's greatest mathematicians. Most of the book explains Hardy's criteria for beauty and profundity in mathematics. However, a long foreword by C. P. Snow expands upon Hardy's terse information with anecdotes about their friendship and years as colleagues at Cambridge. Both Hardy's apology and Snow's foreword are dexterously written, witty, and clear in explaining mathematical ideas. 153 pp.

Hardy, Godfrey Harold. *Ramanujan*. Cambridge, England: Cambridge University Press, 1940; New York: Chelsea Publishing, 1978.
Scientific biography. A biography and explication by a great

mathematician, G. H. Hardy, about an even greater mathematicians, Srinivasa Ramanujan (1881-1920). Based on twelve lectures, Hardy's book opens with a brief biography of Ramanujan and then dives into the intricate mathematical advances he made, including his theory of prime numbers, analytic theory of numbers, and functions. Best appreciated by mathematicians. Photos; bibliography; end notes for each chapter. 236 pp.

Heims, Steve J. *John Von Neumann and Norbert Wiener.* Cambridge, Mass.: MIT Press, 1980.
Biography. Heims wants to show the basic change in the culture of science and people's attitude toward it that occurred, he believes, after atomic bombs were dropped on Japan during World War II. He uses John von Neumann (1903-1957) and Norbert Wiener (1884-1964) as foci of his study because both made names as mathematical prodigies before this watershed and both participated in weapons development after it. Heims admits his bias against nuclear weapons but insists his study still reveals a radical change in American culture. He includes biographical information about the youth, education, and overall careers of his two subjects but naturally concentrates on their post-World War II work. Photos; end notes; index. 547 pp.

Hodges, Andrew. *Alan Turing: The Enigma.* New York: Simon and Schuster, 1983.
Biography. Hodges carefully unfolds the education, personal life, career, and ideas of Alan Turing (1912-1954). An applied mathematician, Turing helped unlock the secret of the Germans' encoding device, the "enigma machine," during World War II and laid the foundations for computer science. He also devised early means of distinguishing computer operations from human intelligence—the Turing test. Hodges writes for general readers. Photos and drawings; end notes; index. 587 pp.

Hofmann, Joseph E. *Leibniz in Paris 1672-1676.* Cambridge, England: Cambridge University Press, 1974.
Scientific biography. The author closely explains the devel-

opment of Gottfried Wilhelm von Leibniz (1646-1716) into a mature, creative mathematician. The text is highly technical, requiring a sophisticated understanding of mathematics. Hofmann discusses the invention of calculus and the quarrel with Isaac Newton over priority. Bibliography; index. 372 pp.

Infeld, Leopold. *Whom the Gods Love: The Story of Évariste Galois*. New York: Whittlesey House, 1948; Reston, Va.: National Council of Teachers of Mathematics, 1978.
Biography. Using fictional techniques, such as dialogue, Infeld presents a dramatic version of the life of Évariste Galois (1811-1832). Galois died in a duel following brilliantly innovative work on solving equations with radicals and prime numbers. His short life was colorful, and Infeld labors to make it read like a novel. The reader should beware that he is trying to capture the essence of Galois' life and the cultural atmosphere of nineteenth-century France rather than document historical facts. Bibliography. 323 pp.

Kac, Mark. *Enigmas of Chance*. New York: Harper and Row, 1985.
Autobiography. Both charmingly written and fairly technical in its mathematics, this book reveals a thoughtful man in love with his career. Kac pioneered probability theory, encouraging students and other scientists to distrust axiomatic thinking. He also helped develop statistics, number theory, and the stochastic modeling of differential equations. His purpose in writing, he says, is to show how he became a mathematician and reveal the rich inner life it afforded him. Readers unfamiliar with mathematics may find the equations, graphs, and references to other mathematicians daunting. Diagrams; index. 163 pp.

Kanigel, Robert. *The Man Who Knew Infinity: A Life of the Genius Ramanujan*. New York: Charles Scribner's Sons, 1991.
Biography. The short life of Srinivasa Ramanujan from Madras, India, is one of the most remarkable success and tragedy stories in the history of science. To Kanigel, in this finely written book, it is a story of an "inscrutable intellect and a simple heart." Ramanujan's breathtaking innovations in higher mathematics are still regarded with awe, and some of

his ideas have found application in chemistry and computer science, as Kanigel explains. The story also involves a great friendship and collaboration with G. H. Hardy and J. E. Littlewood, the Cambridge mathematicians who went to great lengths to bring Ramanujan to England and secure his place in the world of mathematics. Readers will need a college-level knowledge of math to appreciate the text fully, although most of it is accessible to general readers. Photos; end notes; index. 438 pp.

Kennedy, Don H. *Little Sparrow: A Portrait of Sophia Kovalevsky*. Athens: Ohio University Press, 1983.
Biography. Kennedy is less interested in Sophia Kovalevsky (1850-1891) as a scientist than as an unusual woman who outgrew her origins in Russia's landed gentry to become independent and self-fulfilled. Kennedy considers her to have been a transitional figure between the old Russia of Czarist times and the Russia of the Soviet Union. He treats her mathematical work only superficially. Photos and drawings; bibliography; index; 341 pp.

Kennedy, Hubert C. *Peano*. Dordrecht, Holland: D. Reidel Publishing, 1980.
Biography. Giuseppe Peano (1858-1932) was a leading Italian mathematician in the early twentieth century, best known for his postulates about natural numbers and his space-filling curve. But with more than 200 publications, he ranged widely in mathematics, and Kennedy writes to capture the range and depth of his research, as well as the details of his life and academic career. The story contains glimpses of long-dead fads of the intelligentsia, such as an artificial international language (Peano propose one based on Latin), and controversy, especially his quarrel with Vito Volterra about how a cat turns legs-down while falling. The text contains moderately sophisticated mathematics. Photo; bibliography; indexes. 230 pp.

Koblitz, Ann Hibner. *A Convergence of Lives*. Boston, Mass.: Birkhäuser, 1983.
Biography. As Koblitz explains in her preface, Russian science entered a golden age in the second half of the nineteenth cen-

tury, a time when the intelligentsia looked forward to social reform and equality for women. Sophia Kovalevsky grew up in this era, and the freethinking scientific community of Europe recognized her genius for mathematics and helped her develop it. Koblitz emphasizes the importance of the times and the international scientific community for Kovalevsky and explains how mathematics was a fundamental part of her life. For general readers. Photos and diagrams; bibliography; index. 305 pp.

Kovalevskaya, Sofya. *A Russian Childhood*. New York: Springer-Verlag, 1978.
Autobiography. One of the first women to win a professorship as a mathematician (in Sweden), Sophia Kovalevsky looks back on her youth in Russia and life among the country's gentry. She recounts the source of her interest in mathematics, her education at the University of Berlin, and, in a separate essay, her career as a mathematician. The volume concludes with a long essay by P. Y. Polubarinova-Kochina that analyzes Kovalevasky's work in physics and mathematics. A charming book, although the concluding essay requires a sophisticated knowledge of mathematics to understand. Brief bibliography; notes at chapter ends. 250 pp.

Lenzen, Victor F. *Benjamin Peirce and the U.S. Coast Survey*. San Francisco, Calif.: San Francisco Press, 1968.
Biography. Lenzen's short book describes the career of the early nineteenth-century astronomer, mathematician, and superintendent of the U.S. Coast Survey, Benjamin Peirce (1809-1880). Although he conducted exacting measurements of the planets Neptune and Uranus, he is best known for his survey work, a crucial job for the young United States. Photos; index. 54 pp.

Lützen, Jesper. *Joseph Liouville, 1808-1882: Master of Pure and Applied Mathemtics*. New York: Springer-Verlag, 1990.
Scientific biography. The author's aim is to recount Joseph Liouville's scientific career; his personal life receives less attention. The first six chapters relate that career in chronological order. The remaining ten chapters deal with his math-

ematical work, which included fractional calculus, integration, eigenvalues, and potential theory. He also explored electrodynamics and mechanics. A specialist's knowledge of mathematics is required to understand the second half of the book. Photos and drawings; bibliography; index. 884 pp.

MacHale, Desmond. *George Boole*. Dublin, Ireland: Boole Press, 1985.
Biography. MacHale presents a full biography of George Boole (1815-1864), considering his achievements in mathematics, his personal life, and his academic career. An Englishman transplanted to County Cork, Ireland, where he taught at University College, Boole developed a mathematical language for logic, since known as Boolean algebra. More than a monument to abstract thinking, Boolean algebra underlies circuitry and software for modern computers and computer systems, including the Internet. Boole himself was keenly interested in the applications of his ideas. He also wrote poetry, some of which MacHale prints, and threw himself into religious and social controversies. MacHale provides mathematical demonstrations of Boole's discoveries, which require a knowledge of advanced algebra, but these demonstrations occupy a small portion of the book. Photos and drawings; end notes; bibliography; index. 304 pp.

Macrae, Norman. *John von Neumann*. New York: Pantheon, 1992.
Biography. Macrae argues that John von Neumann, one of the premier mathematicians of the twentieth century, was a brilliant and dispassionate logician whose clarity of mind extended to his political activity. Specifically, von Neumann helped develop the fission and fusion bombs and counseled U.S. presidents to take a tough stand against Stalinist Russia. Macrae defends the Hungarian-born von Neumann from those who consider him a warmonger and stealer of ideas. His thesis is that von Neumann lacked the revolutionary creativity of such thinkers as Albert Einstein but did have a genius for improving the ideas of others. Those ideas include matters in quantum mechanics, set theory, game theory, and the logic of digital computers. Photo of von Neumann; bibliography; end notes; index. 405 pp.

Masani, Pesi R. *Norbert Wiener, 1884-1964*. Basel, Switzerland: Birkhäuser, 1990.
Scientific biography. Masani studies the colorful and varied career of Norbert Wiener, one of America's great mathematicians. He contributed to cybernetics (his invention), statistical physics, and computer science. He also published his ideas on economics, foreign policy, and education. Although non-specialists will understand much of this biography, Masani goes into depths of technical detail on some topics that require a thorough foundation in mathematics, physics, and computer design. Photos and diagrams; end notes; bibliography. 416 pp.

Monastyrsky, Michael. *Riemann, Topology, and Physics*. Boston, Mass.: Birkhäuser, 1987.
Scientific biography. The pioneering work on topology by Bernhard Riemann (1826-1866) proved unexpectedly important to modern physics, the general theory of relativity in particular. According to Freeman Dyson in his introduction to this volume, Riemann's work illustrates the strong power of abstract mathematics to prepare the way for discoveries in other sciences. In the first half of the book, Monastyrsky recounts Riemann's life and career succinctly but with considerable technical detail. The second half of the book explains the use of topology in today's physics. Photos and drawings; bibliography; index. 158 pp.

Moorehead, Caroline. *Bertrand Russell: A Life*. New York: Viking, 1992.
Biography. Bertrand Russell (1872-1970), although best known as a philosopher, popular writer, and political reformer, was a leading mathematician, logician, and philosopher of science at the beginning of the twentieth century. *Principia Mathematica*, which he wrote with Alfred North Whitehead, attempted to find a single logical foundation for mathematics, an ambitious undertaking later obviated by Kurt Gödel's incompleteness theorem. In this engrossing biography, Moorehead is more interested in Russell's personal struggles—especially his relations with women—and politics than his work in science; nonetheless, her summary of his mathematical work, and his famous paradox about classifications, is lucid and set firmly in

the larger context of his intellectual pursuits. For general readers who enjoy philosophy. Photos; bibliography; index. 596 pp.

O'Donnell, Seán. *William Rowan Hamilton*. Dublin, Ireland: Boole Press, 1983.
Biography. O'Donnell considers William Rowan Hamilton (1805-1865) Ireland's greatest mathematician. During Hamilton's youth, his contemporaries saw in him a second Isaac Newton, and although most now do not agree, Hamilton's mathematical discoveries were in fact extensive and important to later physics, quantum mechanics and the theories of relativity in particular. O'Donnell describes him as a genius who fits the stereotype of the absent-minded professor, who was a failure as an astronomer, although he was Ireland's Astronomer Royal, and who had quixotic ambitions to make his everlasting reputation as a poet—a fascinating character, in short. Accordingly, O'Donnell concentrates on his upbringing, scientific career, and personality, but he also carefully explains Hamilton's work on advanced algebra and optics. Photos and diagrams; end notes; index. 224 pp.

Ore, Oystein. *Niels Henrik Abel, Mathematician Extraordinary*. Minneapolis: University of Minnesota Press, 1957.
Biography. The Norwegian mathematician Niels Henrik Abel (1802-1829) had a brief but brilliant career before dying of tuberculosis. His work on fifth-order equations and transcendental functions caught the attention of the eminent mathematicians of the nineteenth century. Ore writes reverently of Abel in this exhaustively researched book, which is suitable for general readers interested in the history of mathematics. Photos and drawings; bibliography; index. 277 pp.

Poundstone, William. *Prisoner's Dilemma*. New York: Doubleday, 1992.
Biography-popular mathematics. Poundstone devotes about half this book to John von Neumann's life and career in mathematics, academia, the Manhattan Project, consulting for the RAND Corporation and other companies, and service on the Atomic Energy Commission. The real center of the book,

however, is less von Neumann than his most famous mathematical creation, game theory. The prisoner's dilemma is the field's most celebrated problem, whose essence and variations Poundstone explains in extensive detail and whose applications in economics and military strategy he also discusses at length. Along the way, he explains why von Neumann was so vociferous a proponent of nuclear armament, even "preventive" nuclear war against the Soviet Union, following World War II. Unfortunately, the biographical sections are not nearly as well written as the math explication. Poundstone dwells on von Neumann's amazing memory and calculating skills but seems on less sure ground talking about von Neumann's personal life. Photos and diagrams; bibliography; index. 294 pp.

Reid, Constance. *Courant in Göttingen and New York: The Story of an Improbable Mathematician*. New York: Springer-Verlag, 1976. Biography. Richard Courant (1888-1972) was one of the many scientific leaders who had to flee the Nazi regime in the 1930's. Before that he had established himself as a premier mathematician, the intellectual heir of Felix Klein and David Hilbert at the University of Göttingen. He moved to New York University and was soon a leading American mathematician as well, founder of the Courant Institute of Mathematical Sciences. He was a controversial figure, as Reid illustrates abundantly. Yet few question the greatness of his mathematics, which included work on the eigenvalues of differential equations, and he worked with or knew many of the preeminent mathematicians of the twentieth century. Based on interviews with Courant, Reid's biography is suited for a general audience. Photos; index. 314 pp.

Reid, Constance. *Hilbert*. New York: Springer-Verlag, 1970. Biography. Widely regarded as among the greatest twentieth-century mathematicians, David Hilbert (1862-1943) made major contributions to modern algebra, geometry, analysis, mathematical physics, and the philosophy of mathematics. In fact, his ideas and program for mathematics in general long dominated the discipline. This dominance came partly from his position as the leading spirit of the mathematics department at the University of Göttingen, one of the most fertile groups ever, and partly from a speech he delivered in 1900 about the course of twentieth-century research, which named twenty-three fundamental problems that he thought should be studied. Reid describes the creative variety of Hilbert's career in detail but with the focus on the essential ideas rather than on mathematical techniques. She also discusses his private life in passing. An essay by Hermann Weyl (Hilbert's student and friend) outlines the technical achievements in Hilbert's work. Photos; index. 290 pp.

Reid, Constance. *Neyman—from Life*. New York: Springer-Verlag, 1982.
Biography. Collaborating with Egon Pearson, Jerzy Neyman (1894-1981) put together the foundations of modern statistics. Reid's biography of Neyman rests on her conversations with him and reminiscences of his colleagues. The account is nontechnical and jumps rapidly among the periods of his life, from his youth in Ukraine to his tenure as professor at the University of California, Berkeley. Photos; index. 298 pp.

Reid, Constance. *The Search for E. T. Bell*. Washington, D.C.: Mathematical Association of America, 1993.
Biography. Born in Scotland, Eric Temple Bell (1883-1960) studied mathematics at Stanford University and later became a professor at the California Institute of Technology, where he earned famed both as an eccentric and as a number theorist. He also wrote a widely read popular mathematics book, *Men of Mathematics*, and, under the pseudonym John Taine, sixteen science fiction novels. Reid's book is both a biography and the story of her research, which took quirky turns, for Bell and his family led unusual, sometimes mysterious lives whose details were difficult to track down. A lively book for a general audience. Photos; index. 372 pp.

Stein, Dorothy. *Ada: A Life and a Legacy*. Cambridge, Mass.: MIT Press, 1985.
Biography. According to Stein, Augusta Ada Byron, Countess of Lovelace (1815-1852), was a romantic, mysterious figure during her times. The daughter of the poet Byron perhaps

could not help being so. Her friendship and work with Charles Babbage, who designed a precursor to the computer, has brought Byron fame to Computer Age readers, yet some claims about her contributions are fanciful, Stein says. Stein, a former computer programmer herself, scrutinizes Byron's programming for Babbage's computing machines. The entire biography, in fact, reflects careful reviews of original documents and judicious analysis of the evidence about her life. Photos and drawings; end notes; index. 321 pp.

Stigt, Walter P. van. *Brouwer's Intuitionism.* Amsterdam, The Netherlands: North-Holland, 1990.
Scientific biography. Luitzen Egbertus Jan Brouwer (1881-1966) contributed basic ideas in topology, set theory, and logic. The intellectual foe of David Hilbert, Brouwer was a combative rebel who saw his life work as the reform of mathematics based on his intuitionist philosophy, which eschewed objective manipulation of symbols for its own sake and held the individual mind the source of mathematical activity. Stigt's book starts with two chapters of biographical material; the remaining four chapters discuss in great technical detail Brouwer's philosophy and mathematical achievements, and these chapters require a sophisticated understanding of advanced mathematics and logic to comprehend. Photos; bibliography; index. 530 pp.

Ulam, S. M. *Adventures of a Mathematician.* New York: Charles Scribner's Sons, 1976.
Autobiography. A loosely connected series of anecdotes mixed with some philosophical reflections, the book is odd in that in talks much about abstruse mathematics, especially topology and number theory, but it does not contain a single formula or calculation. Instead, Ulam recalls the many prominent mathematicians and physicists he worked with in his long career. Because he worked on the Manhattan Project, the hydrogen bomb, and the development of computers, he knew a great many: Richard Feynman, John von Neumann, Edward Teller, Enrico Fermi, and Norbert Wiener, among many others. Ulam regularly passes judgment on their overall intelligence as well as they ability in mathematics; his opinions

sometimes vary strikingly from the person's public reputation. Photos; short bibliography; index. 317 pp.

Wang, Hao. *Reflections on Kurt Gödel*. Cambridge, Mass.: MIT Press, 1988.
Biography. Kurt Gödel is best remembered for his proof that no mathematical system can prove its own completeness, a disturbing finding to his colleagues. But as Wang ably shows, Gödel interested himself in theoretical physics, in part at least because of his friendship with Albert Einstein, and philosophy. Wang's treatment is sophisticated, although readers do not need to be specialists in logic, philosophy, or mathematics to follow the text. Photos; index. 336 pp.

Weil, André. *The Apprenticeship of a Mathematician*. Basel, Switzerland: Birkhäuser, 1992.
Autobiography. Largely avoiding the technicalities of mathematics, Weil (b. 1906) reminisces about his career. Educated in France, he became a professor at universities in Germany, India, Brazil, and the United States, finishing his career at the Institute for Advance Studies. Truly an international mathematician, Weil contributed to both advanced mathematics and mathematical physics and was a founder of the influential Bourbaki Group of mathematicians. His sister was the famous radical philosopher Simone Weil, although she does not figure prominently in this book after their youth. Since Weil knew and worked with many of the most famous people in modern mathematics and physics, his book is a valuable resource for science history, especially the turbulent times during the world wars. Photos; index. 197 pp.

Wiener, Norbert. *Ex-Prodigy: My Childhood and Youth*. New York: Simon and Schuster, 1953.
Autobiography. To say that this biography reveals a man full of himself is an incalculable understatement. The tone is haughty, the style pompous, and the theme, largely, self-congratulations. Wiener was in fact a genius in mathematics and the inventor of cybernetics. As a prominent professor at the Massachusetts Institute of Technology, he knew many leading mathematicians and scientists of the mid-twentieth century.

Even if his version of events is biased, sometimes stridently so, this book affords insights into the temper of the times and a perversely fascinating view of a brilliant thinker. Photos; index. 309 pp.

Wiener, Norbert. *I Am a Mathematician: The Later Life of a Prodigy.* Cambridge, Mass.: MIT Press, 1956.
Autobiography. Wiener continues the story of his life in this second volume from the time the Massachusetts Institute of Technology hired him. During his long career he worked with, and sometimes became embroiled in controversy with, many of the great minds of European and American mathematics, including John von Neumann. Wiener explains one of his central interests, cybernetics, in detail for general readers. Index. 380 pp.

Chapter 7

Medical Sciences

COLLECTIONS

Bindman, Lynn, Alison Brading, and Tilli Tansey. *Women Physiologists*. London: Portland Press, 1993.
Tansey opens this collection of profiles with an essay about women in the British Physiological Society, the book's sponsor. Then other women physiologists write about their illustrious forebears. They are 18 in all, arranged in chronological order by birth year, beginning with Dame Hariette Chick (1875-1977) and ending with Brenda Muriel Schofield (1926-1968). Index. 161 pp.

Clark, Paul F. *Pioneer Microbiologists of America*. Madison: University of Wisconsin Press, 1961.
Biography-history. Clark chronicles the history of microbiology in American by profiling its leading practitioners and their discoveries. After summarizing the state of knowledge in the early eighteenth century, he starts his survey with the attempt by Cotton Mather and Zabdiel Boylston to stem a 1721 smallpox epidemic in Boston by inoculating volunteers. He continues with discussions of early sanitation, the role of Johns Hopkins Medical School, agricultural experiment stations, and the contribution of federal agencies and research institutes to microbiology up until about 1920. Finally, he sketches the careers of microbiologists by regions of the United States, mentioning dozens of them. Although full of information, this is not a book suited to quick reference. Photos; end notes; index. 369 pp.

Crowther, J. G. *Six Great Doctors*. London: Hamish Hamilton, 1957.
Crowther supplies simply written, sympathetic biographical sketches of six scientists. Only two were really physicians, William Harvey (1578-1657) and Ronald Ross (1857-1932), who discovered how to prevent malaria. The rest contributed some innovation to medicine: Louis Pasteur (1822-1895) and the germ theory of disease; Joseph Lister (1827-1912) and antiseptic surgery; Ivan Petrovich Pavlov (1849-1936) and conditioned reflex; and Alexander Fleming (1881-1955) and penicillin. Photos and portraits; brief bibliographies for each chapter. 207 pp.

Curtis, Robert H. *Great Lives: Medicine*. New York: Charles Scribner's Sons, 1993.
For readers 12 years and older. Curtis tells his readers that everyone, just by being born, contributes to the world of science because each person becomes part of the vital statistics that medical science evaluates in order to reach conclusions about disease. The 38 scientists he profiles, from Hippocrates (460-370 B.C.E) to Joseph E. Murray (b. 1919), all illustrate the search for knowledge and application of techniques in medical sciences. Curtis arranges them in handy categories: ancient medicine; beginnings of modern medicine; drugs; medical devices; infectious diseases; medicine as art and skill; mental health; and modern times and cures. Curtis writes clearly about both the historical background and the scientific principles. Photos and drawings; bibliography; index. 326 pp.

De Kruif, Paul. *Men Against Death*. New York: Harcourt, Brace, 1932.
One of De Kruif's most compelling books of popular science, this volume opens with a long essay about life expectancy and the helplessness of physicians to improve it much until the early twentieth century. Then De Kruif profiles the leaders in medical research and bacteriology: Ignac Philipp Semmelweis, Frederick Banting, George R. Minot, R. R. Spencer, Fritz Schaudinn, Jules Bordet, Julius Wagner-Jauregg, Niels Finsen, Auguste Rollier, and Ove Strandberg. The profiles concern only the research careers of these men. The medical principles

receive clear, if often slangy, explanation for general readers, and one gets a definite sense of the subjects' personalities. The book's remarkable quality is the excitement and sense of wonder that scientists were becoming able to prevent or cure diseases that had commonly been killers before. Photos; index. 363 pp.

Flexner, James Thomas. *Doctors on Horseback*. New York: Viking, 1937.
As Flexner puts it, the early doctors of America were part of every adventure in extending European civilization across the American continent and in learning how to deal with new diseases and old in the wilderness. He profiles seven of these pioneers: John Morgan (1735-1789); Benjamin Rush (1745-1813); Ephraim McDowell (1771-1830); Daniel Drake (1785-1852); William Beaumont (1785-1853); Crawford W. Long (1815-1878); and William T. G. Morton (1819-1868). The text, for a general audience, is pleasant reading, an informative introduction to the struggle to cope with medical needs in America. Photos and portraits; bibliography; index. 370 pp.

Fox, Daniel M., Marcia Meldrum, and Ira Rezak, eds. *Nobel Laureates in Medicine or Physiology*. New York: Garland Publishing, 1990.
The alphabetized entries extend from 1901 (Emil von Behring) to 1989 (J. Michael Bishop and Harold Elliot Varmus), 149 scientists. Written by university scholars, the entries are directed at general readers and summarize the winner's career in about 2,000 to 2,500 words. A bibliography concludes each entry, containing works by both the featured scientist and secondary sources. Index. 595 pp.

Hendin, David. *The Life Givers*. New York: William Morrow, 1976.
Every modern American, Hendin tells his readers, has been affected by at least one of the six doctors whom he profiles in this entertaining book. They are Jonas Salk, developer of the polio vaccine; Howard Rusk, who pioneered rehabilitating the crippled; C. Walton Lillehei, who developed open heart surgery; Irving Cooper, a neurosurgeon who devised a proce-

dure to relieve the palsy of Parkinson's disease; Nathan Kline, who invented drugs to treat mental illness; and John Rock, one of the fathers of the birth control pill. Hendin's style is light—he often uses dramatic dialogue and humor—but his focus on the achievements of these men is clear and serious. Index. 260 pp.

Jirka, Frank J. *American Doctors of Destiny*. Chicago, Ill.: Normanie House, 1940.
The grandiose title of this book matches its grandiose prose, yet it features 19 biographical portraits of influential American physicians by a physician, and for some of them popular accounts are difficult to find. Among the subjects are Benjamin Rush, Samuel Mudd, Oliver Wendell Holmes, Walter Reed, and William Worrell Mayo. For readers who do not mind the sometimes inflated style, there is much information about the careers and medical ideas of the subjects. Portraits; bibliography; index. 361 pp.

Kaufman, Sharon R. *The Healer's Tale: Transforming Medicine and Culture*. Madison: University of Wisconsin Press, 1993.
Biography-history. Kaufman implements an intriguing idea: to illustrate the changes in medicine during the twentieth century with a combination of the life stories and reminiscences of seven leading practitioners. The seven are J. Dunbar Shields, Saul Jarcho, Paul Bruce Beeson, Mary B. Olney, Jonathan Evans Rhoads, C. Paul Hodgkinson, and John Romano. They trained as doctors during the first third of the century and saw medicine become more specialized, powerful, technology intensive, and expensive. Kaufman considers the ethical problems involved with those changes as well as the cultural influence of such physicians as her seven. To better emphasize the evolution of medicine, she divides the information about each subject into four sections: training to become a doctor during the 1920's; concentrating on a specialty during the 1930's; applying the increasing power to cure diseases, thanks to pharmaceutical research, during the 1930's and 1940's; and overseeing the expanding power of physicians from 1946 into the 1970's. Photos; end notes; index. 354 pp.

Lambert, Samuel W., and George M. Goodwin. *Medical Leaders.* Indianapolis, Ind.: Bobbs-Merrill, 1929.

Following an introduction on the history of medicine, the authors provide a chronologically arranged series of biographical sketches. They begin with Hippocrates (c. 460-370 B.C.E.) and finish with William Osler (1849-1919); between are 20 of the greats of medicine. The sketches are brief, addressed to a general audience, and separated by short essays supplying historical or scientific background. Photos and drawings. 331 pp.

McKown, Robin. *Heroic Nurses.* New York: G. P. Putnam's Sons, 1966.

After an introduction concerning the history of nursing, McKown profiles the lives, good deeds, and careers of 12 nurses from around the world. Arranged in chronological order from France's Jeanne Mance (1606-1673) to Sister Dulce Lopes Pontes in modern Brazil, the list includes such familiar names as Florence Nightingale and Clara Barton (1821-1912), as well as some unexpected names, such as Nathaniel Hawthorne's daughter Rose Hawthorne Lathrop and the daughter of Emperor Haile Selassie of Ethiopia, Princess Tsahai Haile Selassie. The text is suitable for high school students; among McKown's purposes in writing is to inspire nursing careers. Photos; bibliography; index. 320 pp.

Marks, Geoffrey, and William K. Beatty. *Women in White.* New York: Charles Scribner's Sons, 1972.

The authors discuss the role of women in medicine throughout the ages. Their first section concerns women healers, some legendary, in the ancient and medieval world, especially midwives. The second section recounts the careers of 15 women who struggled their way into medicine despite biases against them; among them are Elizabeth Garrett Anderson (1836-1917), Emily Dunning Barringer (1876-1961), and Alice Hamilton (1869-1970). The final section includes women whose work was related to medicine, even though they were not physicians, such as Florence Nightingale (1820-1910) and Marie Curie (1867-1934), who used X-rays to help treat the wounded during World War I. In an epilogue, the authors discuss women in modern medicine. Photos; bibliography; end notes; index. 239 pp.

Metchnikoff, Elie. *The Founders of Modern Medicine*. New York: Walden Publications, 1939.
The author writes at length of Robert Koch, Joseph Lister, and Louis Pasteur. The style is both personal and professorial in tone. Metchnikoff provides generous discussions of the scientific discoveries of his subjects, although some allowances must be made for his technical explanations, because he wrote in the 1930's. 387 pp.

Nobel Prize Winners: Physiology or Medicine. 3 vols. Pasadena, Calif.: Salem Press, 1991.
Includes Nobel Prize winners from 1901 (Emil von Behring) to 1990 (Donnall E. Thomas and Joseph E. Murray), 151 scientists altogether. Following an extensive introductory essay about the prize for medicine or physiology, the entries, chronologically arranged, have five sections: short list of vital statistics; summary of the award ceremony and Nobel lecture; personal biography; scientific career; and annotated bibliography. Each entry is about 3,500 words long, and the greatest portion concerns the winner's scientific career, which is recounted in detail for general readers. Photos; index in the third volume.

Nuland, Sherwin B. *Doctors: The Biography of Medicine*. New York: Alfred A. Knopf, 1988.
Biography-history. A well-written, absorbing book, it relates the history of Western medicine by profiling the most influential and innovative thinker in each major era. The featured doctors are Hippocrates, Galen, Andreas Vesalius, Ambroise Paré, William Harvey, Giovanni Morgagni, John Hunter, René Laennec, Ignac Semmelweis, Rudolf Virchow, Joseph Lister, William Stewart Halsted, and Helen Taussig. Nuland admits that the selection is somewhat idiosyncratic, and in the narrative he frequently refers to his own experience as a doctor and biographer and editorializes about some historical issues in medicine. Chapters also describe the origins of general anesthesia and organ transplantation. Photos and drawings; extensive bibliography; index. 518 pp.

O'Hern, Elizabeth Moot. *Profiles of Pioneer Women Scientists.* Washington, D.C.: Acropolis Books, 1985.

O'Hern offers profiles of 20 women microbiologists and public health workers, most of whom flourished in the mid-twentieth century. O'Hern stresses the difficulty these women had in winning equal status with male colleagues, even when working in the forefront of medical research on diseases and employed in government agencies. The profiles are organized into three sections: New York and vicinity; the National Institutes of Health; and the West Coast, Midlands, and elsewhere. Among the subjects are Florence Rena Sabin (1871-1953); Leona Baumgartner (b. 1902), first woman health commission of New York City; Bernice Elaine Eddy (b. 1903); and Elizabeth McCoy (1903-1978). Photos; end notes; glossary; index. 264 pp.

Sourkes, Theodore L. *Nobel Prize Winners in Medicine and Physiology 1901-1965.* New York: Abelard-Schuman, 1966.

An updated edition of Lloyd G. Stevenson's book (see entry below), adding 31 more scientists and following the same format. Photos; index. 464 pp.

Stevenson, Lloyd G. *Nobel Prize Winners in Medicine and Physiology 1901-1950.* New York: Henry Schuman, 1953.

The coverage, from Emil von Behring in 1901 to Edward Calvin Kendall, Philip Showalter Hench, and Tadeus Reichstein in 1950, includes 60 scientists. Each entry offers a biographical sketch , a description of the prize-winning work, the consequences of that work in theory and practice, and, in some cases, a brief bibliography—about 1,500 words in all. Most of the explanations of the scientific work are quoted from other writers, making the text a stylistic hodgepodge. Chapters lump together cowinners of a year's prize, a major disadvantage in that less space for information is allotted to each to maintain the uniform chapter length. Photos; index. 291 pp.

Thacher, James. *American Medical Biography.* 2 vols. New York: Da Capo Press, 1967.

Of interest primarily to historians, this is a reprint of a collection of profiles published in 1828. Thacher begins with a history of medicine up to the time of publication. Then come the profiles, most of them a few pages in length, but some of them substantial, such as those of John Jones and Benjamin Rush. Thacher is not often specific about the details of medical treatment but supplies abundant detail about the conduct of medical practice and its political environment. Portraits; bibliography in vol. 2.

Walsh, James Joseph. *Makers of Modern Medicine*. Freeport, N.Y.: Books for Libraries Press, 1970.
Biography-history. In this reprint of a 1907 edition, Walsh provides an introduction about the rise of modern medicine and then discusses the scientific careers of key nineteenth-century physicians and researchers: John Baptist Morgagni, Leopold Auenbrugger, Edward Jenner, Aloysius Galvani, René Theodore Laennec, Robert Graves, William Stokes, Dominic Corrigan, Johann Müller, Theodore Schwann, Claude Bernard, Louis Pasteur, and Joseph O'Dwyer. His intention is to trace major ideas in the rise of medicine, and he uses biographical information to serve that purpose. One photo of Pasteur; index. 362 pp.

BIOGRAPHIES AND AUTOBIOGRAPHIES

Babkin, B. P. *Pavlov*. Chicago, Ill.: University of Chicago Press, 1949.
Biography. Best known for his studies of conditioned reflexes in dogs, Ivan Petrovich Pavlov (1849-1936) earned fame first for perfecting a technique for surgically creating a gastric fistula in dogs and then for his studies of digestive action and the role of the vagus nerve. The studies of digestion earned him the 1904 Nobel Prize for Physiology or Medicine. After the creation of the Soviet Union, Pavlov, already an elder statesman of science, was touted as a model modern scientist and for a while had great influence on Russian research. The author, "Pavlov's senior surviving pupil," devotes the first part of the book to his mentor's life and career. The following three parts explain his physiological research of the heart,

digestive glands, and conditioned reflexes. Photo of Pavlov; bibliography; end notes; index. 365 pp.

Benison, Saul. *Tom Rivers: Reflection on a Life in Medicine and Science*. Cambridge, Mass.: MIT Press, 1967.
Biography-history. A resource for scholars, this volume contains a question-and-answer format interview with Tom Rivers (1889-1962). Rivers conducted research in viruses and administered research programs. He worked at the Rockefeller Institute, National Foundation for Infantile Paralysis, and Public Health Research Institute of New York. In the text, he reminisces about his professional life and acquaintance with the leading microbiologists of the United States. Photos; bibliography; glossary; index. 682 pp.

Benison, Saul, A. Clifford Barger, and Elin L. Wolfe. *Walter B. Cannon: The Life and Times of a Young Scientist*. Cambridge, Mass.: Belknap Press, 1987.
Biography. Walter B. Cannon (1871-1945) was a physiologist who investigated the digestive tract and adrenal gland and pioneered the use of X-rays for medical research. He also helped place the Harvard Medical School on a sound scientific basis and dealt with controversies involving medical research, especially the conflict with antivivisectionists. His life affords a view of American academic medical research at its most advanced in the first decade of the twentieth century, and the authors offer readers that view in a well-written scholarly work. Specifically, they focus on his life and career up to 1917. Photos; bibliographical note; end notes; index. 520 pp.

Brock, Thomas D. *Robert Koch, a Life in Medicine and Bacteriology*. Madison, Wis.: Science Tech Publishers, 1988.
Biography. Robert Koch (1843-1910) founded bacteriology, not only through his work on the anthrax bacillus and his discovery of the tubercle bacillus but also by championing the germ theory of disease. His view, now taken for granted, was revolutionary in the nineteenth century. Considering Koch's importance and his Nobel Prize for Physiology or Medicine in 1905, it is strange that Brock's book is the first analytical, full-length biography in English. Well aware of that fact, Brock

studies Koch's publications closely in explaining his ideas. Brock also relates Koch's childhood, medical education, married life, academic career, travels to Asia and America, and relations with the other prominent germ-theory advocates of the times, including Louis Pasteur, with whom he had a long-running quarrel, and Joseph Lister. Photos and drawings; bibliography; end notes; index. 364 pp.

Brown, Jordan. *Elizabeth Blackwell.* New York: Chelsea House Publishers, 1989.
Biography for readers 12 years and older. Brown opens the book with the scene of the medical school graduation of Elizabeth Blackwell (1821-1910). It was an extraordinary achievement for Blackwell, whom Brown calls the first woman physician of modern times. As Brown shows, Blackwell had to battle biases against educating women to realize her dream of becoming a doctor, and she had to go to Switzerland to do it. Brown discusses feminist efforts toward greater opportunity in the nineteenth century as well as Blackwell's life, attempts to help the poor and slaves, and practice of medicine. Photos and drawings; bibliography; index. 110 pp.

Burnet, Macfarlane. *Changing Patterns: An Atypical Autobiography.* New York: Elsevier Publishing, 1969.
Scientific autobiography. Macfarlane Burnet (1899-1985) shared the 1960 Nobel Prize for Physiology or Medicine with Peter Medawar for research in the microbiology of viruses. That research included bacteriophages, polio, influenza, herpes, and autoimmune disease. He writes on these and other medical topics at a sophisticated level. The book is biographical in that he discusses his development and career and appraises his own work, but it is also a chronicle of the changes in the science of medicine, particularly its growing reliance on studies of pathogens. Photos and diagrams; bibliography; index. 282 pp.

Cannon, Walter Bradford. *The Way of an Investigator.* New York: Hafner Publishing, 1965.
Autobiography. A physiologist at Harvard University for 40

years, Walter Bradford Cannon recalls the colleagues he worked with and the development of his specialty. He describes his youth and background, but the bulk of the book concerns his experiences at Harvard Medical School and as visiting professor in China and elsewhere before World War II. He claims to write for historians and young people considering a career in medicine; his wordy style may discourage less committed readers. Index. 229 pp.

Chauvois, Louis. *William Harvey*. New York: Philosophical Library, 1957.
Biography. According to Chauvois, William Harvey (1578-1657) revolutionized medicine not only in discovering how blood circulates through the body but also in his empirical method of research. That is, he examined the human body directly; he did not rely solely on the writings of the Roman physician Galen, who was held by most other physicians to be the final medical authority. The biography opens with a chapter taking the reader through a day in Harvey's life in 1627, when he was at the height of his intellectual powers. This introduction, dramatically written, strikes a lively tone that carries throughout as Chauvois describes Harvey's education, career, character, published works, research, and professional reputation for general readers. Photos and drawings; bibliography; index. 271 pp.

Clapesattle, Helen. *The Doctors Mayo*. New York: Pocket Books, 1956.
Biography-history. This widely read book follows the careers of three men and the most famous medical name in American history: a father, William Worrall Mayo (1819-1911), and his two sons, William James Mayo (1861-1939) and Charles Horace Mayo (1865-1939), all surgeons. The father maintained a practice in difficult frontier conditions in Minnesota, settling in Rochester. His sons established the world-famous Mayo Clinic there, where many discoveries and innovations were made in medical treatments and surgical techniques; the clinic also inaugurated influential long-term statistical studies. This edition condenses a 1941 edition published by the University of Minnesota. Index. 484 pp.

Cuny, Hilaire. *Ivan Pavlov: The Man and His Theories*. London: Souvenir Press, 1964.
Biography. Cuny allots only the first short chapter of this book to the life of Ivan Petrovich Pavlov, a spare outline that includes his research in digestion. The next four chapters concern aspects of Pavlov's work with conditioned reflexes and his theories of its application to human behavior. The last chapter addresses his Nobel Prize for Physiology or Medicine in 1904 and the reception of his theories. Photos; bibliography; glossary; index. 183 pp.

Cushing, Harvey. *The Life of Sir William Osler*. London: Oxford University Press, 1940.
Biography. By any standard this is a massive, learned tome. Certainly, the life of William Osler (1849-1919) deserves a close examination. Cushing was in a good position to give it; he worked with Osler and knew him well—as his next door neighbor, in fact. Educated in his native Canada and in England, Osler started his career as a professor at McGill University, but his fame began when he moved to Johns Hopkins University. He organized a clinic and then helped open the medical school, which was firmly based upon scientific premises, unlike many American medical schools of the times. Osler's education reforms revolutionized American medical schools after Simon Flexner issued a famous report denouncing practically all the others except Johns Hopkins'. Osler later taught at Oxford University and was knighted. But his work in America and his medical textbook, long a standard, marked the beginnings of modern medicine in this nation. The book, originally published in 1925, is dedicated to medical students, but historians probably will find it most useful, with its detailed discussions of events and generous use of letter excerpts. Photos; footnotes; index. 1417 pp.

Dubos, René J. *The Professor, the Institute, and DNA*. New York: Rockefeller University Press, 1976.
Biography-history. The professor of the title is Oswald Theodore Avery (1877-1955), and the institute is the Rockefeller Institute, where Avery conducted his research and inspired his colleagues. Dubos knew Avery and worked close

by him for 15 years and so brings a firsthand perspective to the topic. Dubos insists that Avery deserved a Nobel Prize for his work on immunity and heredity, especially the role of DNA. Although the Nobel committee considered a nomination for Avery, it failed to act, even though his influence on theoretical biology was great. In the first three chapters, Dubos focuses on the rise of the Rockefeller Institute as a research center and then recounts Avery's personal life, work in the laboratory, research into immunity and bacterial variability, and study of heredity. He concludes with a warm appraisal of Avery's demeanor and philosophy. Although sophisticated in its explanations of biological topics, this book offers a spirited portrait of an early twentieth-century researcher for readers interested in medical history. Photos; bibliography; end notes; index. 238 pp.

Earnest, Ernest. *S. Weir Mitchell, Novelist and Physician.* Philadelphia: University of Pennsylvania Press, 1950.
Biography. In the late nineteenth century S. Weir Mitchell (1828-1914) was as well known for his fiction as for his celebrated rest cure for nervous diseases, his discovery of the nature of rattlesnake venom, and *Gunshot Wounds and Other Injuries of the Nerves* (1864). His friends included Oliver Wendell Holmes, William Osler, and George Meredith. Earnest relates his life story in moderate detail and ponders why Mitchell so quickly fell into obscurity after his death. Photos; end notes; index. 272 pp.

Elders, Joycelyn, and David Chanoff. *Joycelyn Elders, M.D.: From Sharecropper's Daughter to Surgeon General of the United States of America.* New York: William Morrow, 1996.
Autobiography. The first African American woman surgeon general, Elders (b. 1933) left her family farm in rural Arkansas to attend college, the first in her family to do so, and then medical school. She became a professor of pediatric endocrinology at the University of Arkansas, conducting research in thymic hormones. Governor Bill Clinton appointed her to head Arkansas's health department, where she became known for forthright action in protecting children's health and an outspoken manner. She describes her

efforts, foiled in the end, to create national health insurance while in President Clinton's administration. A dramatic story of perseverance and personal success, engrossingly told. Photos; index. 355 pp.

Fisher, Richard B. *Joseph Lister, 1827-1912*. New York: Stein and Day, 1977.
Biography. This close examination of Joseph Lister's life sometimes has a haughty tone but acquaints general readers well with his career as a surgeon and his campaign to introduce antiseptic practices among his colleagues. Before Lister, many post-surgery deaths came from infections, and Lister's contribution to medicine was so fundamental and beneficial that it seems incredible he had to struggle to get his methods accepted. The germ theory of disease was controversial during his career, and he became one of its greatest champions. The story of the theory's triumph comes out as Fisher recounts Lister's family background (at great length), education, professorships in Scotland and London, and public reputation. Photos and drawings; bibliography; index. 351 pp.

Flexner, Simon, and James Thomas Flexner. *William Henry Welch and the Heroic Age of American Medicine*. New York: Viking, 1941; Dover, 1966.
Biography. Co-authored by a famous medical school reformer and director of the Rockefeller Institute, Simon Flexner, this book is a classic of medical biography. In the first quarter of the twentieth century, William Henry Welch (1850-1934) was considered the dean of American medicine. His influence was international influence, and much of this status came from his work in Bellevue Hospital Medical College, where he established the nation's first pathology laboratory, and later at Johns Hopkins University where, with William Osler and Henry Cushing, he started the medical school that soon became the benchmark for excellence in American medical education. At Johns Hopkins he helped base medicine in the scientific method in order to improve the nation's health and bring it up to European standards. The authors tell Welch's story with inside knowledge—the elder Flexner knew and worked with him—explaining why he deserves much of the

credit for modernizing American medical care. Photos; end notes; index. 539 pp.

Franklin, Kenneth J. *William Harvey, Englishman, 1578-1657.* London: MacGibbon and Kee, 1961.
Biography. A distinguished historian of physiology, Franklin emphasizes the extraordinary inventiveness of William Harvey in his studies of blood circulation and anatomy and his practice of medicine and surgery; among his patients was King James I. The style may strike readers as tediously old-fashioned in its frequent use of double-negatives, indirect address and other roundabout phrasing, but the explanations of Harvey's investigations succeed in showing why historians class him as one of the greatest of all medical researchers. Drawings; bibliography; index. 151 pp.

Granit, Ragnar. *Charles Scott Sherrington: An Appraisal.* Garden City, N.Y.: Doubleday, 1967.
Biography. Charles Scott Sherrington (1858-1952) was a professor at Cambridge University and pioneer neurophysiologist. Among the projects of his long career were studies of synapse action and motor control problems. Granit summarizes Sherrington's early life in a loosely structured narrative and then devotes most of the book to explanations of Sherrington's research, teaching, colleagues, and poetry and philosophical works. The chapters on physiological topics assume that readers understand the basic concepts and terminology of the discipline. Photos and diagrams; bibliography; end notes; index. 188 pp.

Gray, Jeffrey A. *Pavlov.* Glasgow, Scotland: Fontana, 1979.
Biography. Gray first discusses the intellectual background and early life of Ivan Petrovich Pavlov in the same short chapter and follows with another chapter, even shorter, about his physiological research. The rest of this brief book summarizes Pavlov's conditioned reflex experiments with dogs, his theory of conditioning and brain function, the nature of personality and psychopathology, and Pavlov's influence on later psychologists. Diagrams; bibliography. 140 pp.

Gronowicz, Antoni. *Béla Schick and the World of Children.* New York: Abelard-Schuman, 1954.

Biography. A pediatrician, Béla Schick (1877-1967) pioneered the study of allergic and immunological responses. He is best known for the scratch test, named after him, for susceptibility to diphtheria. Born in Hungary, Schick trained for medicine in Vienna, Austria, and was a professor in the University of Vienna's children's hospital until he came to the United States in 1923 to practice and conduct research at Mt. Sinai Hospital in New York. His work there made him world famous and brought him numerous medical awards, as the author relates for general readers. The book shows Schick to possess great energy and deep compassion for children. Photos; bibliography; index. 216 pp.

Heiser, Victor. *An American Doctor's Odyssey.* New York: W. W. Norton, 1936.

A minor classic in American autobiography. An adventurer and explorer as well as a doctor, Heiser traveled widely, especially among the Pacific islands. In many places he introduced Western medical practices, and in the Philippines he organized the Bureau of Health, which set sanitation standards. In fact, his remarkable career knew the political extremes of the day: He advised emperors and kings, cared for the Princess of Naples and the Prince of Wales, and ended the incessant warfare between two tribes of head hunters by teaching them to play baseball and take out their aggressions on the diamond. He also conducted research in tropical diseases, including yellow fever and smallpox, and in tuberculosis. A remarkable story throughout. Photo of the author; index. 544 pp.

Horsman, Reginald. *Frontier Doctor: William Beaumont, America's First Great Medical Scientist.* Columbia: University of Missouri Press, 1996.

Biography. William Beaumont (1785-1853) was a U.S. Army surgeon who treated a man wounded in the stomach by a bullet. A gastric fistula formed, allowing Beaumont to see into the man's stomach while it digested food, and Beaumont's experiments and observations revealed many of the stomach's functions and the role of gastric fluids, especially bile. This well-researched scholarly biography traces Beaumont's career in frontier medicine vividly and engagingly. Illustrations; bibliography; index. 320 pp.

Huxley, Elspeth. *Florence Nightingale*. New York: G. P. Putnam's Sons, 1975.
Biography. Huxley studies the forces that blocked Florence Nightingale (1820-1910) from a career in nursing and those that impelled her. Writing for a general adult audience, Huxley tells how Nightingale's proper upper-class rearing and the social demands of the wealthy in Victorian England kept her from hospitals, where doctors and nurses had sordid reputations. But she felt called by God to minister to the sick and wounded. Her chance finally came when the Minister of War allowed her to care for troops in the Crimea. Nightingale not only gained considerable celebrity but also introduced discipline and modern methods to her "Nightingale nurses." A pleasantly written, lavishly illustrated book. Photos and drawings; bibliography; index. 254 pp.

Kass, Amalie M., and Edward H. Kass. *Perfecting the World: The Life and Times of Dr. Thomas Hodgkin, 1798-1866*. New York: Harcourt Brace Jovanovich, 1988.
Biography. Known for the lymph node disease named after him, Thomas Hodgkin was frustrated in his career because he was also a social progressive and thoroughly unconventional. He helped found the University of London and The Aborigines Society, both regarded as subversive, and was a Quaker. Even in his day, he was recognized as an extraordinary physician for his work on aortic valve retroversion, acute appendicitis, and disease prevention. Many of the details of his life have been deliberately obscured, according to the authors, by his colleagues, but they reveal as much of Hodgkin's story as possible, trying always to show the deep complexity and brilliance of his mind. Photos and drawings; end notes; index. 642 pp.

Kean, Benjamin Harrison, with Tracy Dahlby. *M.D.: One Doctor's Adventures Among the Famous and Infamous from the Jungles of Panama to a Park Avenue Practice*. New York: Ballantine, 1990.

Autobiography. "Any doctor's practice provides an unending stream of everything from low comedy to high drama," writes Kean. This book supports the assertion well. A parasitologist, Kean has investigated medical mysteries, some ludicrous and others tragic, in the United States, Europe, Central America, and Africa. He also taught at the Cornell University medical school and maintained a practice in New York City. Among his patients were the Shah of Iran and Salvadore Dali. 402 pp.

Keele, Kenneth D. *William Harvey: The Man, the Physician, and the Scientist*. London: Thomas Nelson and Sons, 1965.
Biography. The title reflects the divisions of this book. The first part summarizes William Harvey's family background, youth, and character. The second part concerns his medical practice, including his lectures and research in anatomy. These sections occupy about a third of the book. The longer final section expands upon Harvey's scientific method, the publication of *De Motu Cordis* (*On the Circulation of Blood*) and its reception, publication of *De Generatione Animalium* (*On the Generation of Animals*), and his legacy to science in general. A readable but plodding exposition. Drawings; bibliography; index. 244 pp.

Kendall, Edward C. *Cortisone*. New York: Charles Scribner's Sons, 1971.
Autobiography. A biochemist, Edward C. Kendall (1886-1972) won the 1950 Nobel Prize for Physiology or Medicine for discovering cortisone. During his career at the Mayo Clinic, he was also involved in developing thyroxin for patients with thyroid secretions deficiency. For a general audience, Kendall recounts his boyhood, education, research efforts, and committee work during World War II for the National Research Council. Photos; index. 175 pp.

Kessler, Henry H. *The Knife Is Not Enough*. New York: W. W. Norton, 1968.
Autobiography. The story of a man with a humanitarian vision, this book describes the rise of rehabilitation as a medical goal by one of its pioneers. Henry H. Kessler was an orthopedic surgeon who first helped industrial accident

victims recover from disabling injuries. Then he extended rehabilitation to all disabled persons, starting his own Kessler Institute in New Jersey. He tells of his work in rehabilitation, which made him world famous, and his experiences in helping the wounded of World War II. Photos; index. 295 pp.

Keynes, Geoffrey. *The Life of William Harvey*. Oxford, England: Clarendon Press, 1966.
Biography. According to Keynes, William Harvey's life is obscure in several important areas. Almost nothing is known of his wife and home life, for instance. Nevertheless, from the scattered sources available Keynes pieces together a picture of Harvey as a person as well as a revolutionary medical researcher. Keynes particularly examines Harvey's relation to his colleagues and patients, among whom were aristocrats and kings. He also describes Harvey's publications and ideas in depth. A scholarly work, the book is also well suited to general readers interested in medical history. Photos and drawings; index. 483 pp.

Kronstadt, Janet. *Florence Sabin*. New York: Chelsea House Publishers, 1990.
Biography for readers 12 years and older. Florence Sabin (1871-1953) gained fame for her research in tuberculosis, inflammation of the lymph nodes, and white blood cells. Her work earned her a professorship at Johns Hopkins University and later a position as researcher at the Rockefeller Institute for Medical Research in New York, both leading centers of medical investigation. She was also the first woman doctor elected to the National Academy of Sciences and received many awards. Kronstadt follows her growth and development into a world famous scientist, from her birth in rural Colorado to the centers of cutting-edge science and science politics. Photos and drawings; bibliography; index. 110 pp.

Lambert, S. M. *A Yankee Doctor in Paradise*. New York: Grosset and Dunlap, 1941.
Memoirs. Sylvester Maxwell Lambert recalls his adventures on South Pacific islands, especially Fiji, as a member of the International Health Board. During the interval between the

two world wars, he helped introduce public health measures to island peoples and studied and treated infectious diseases. Index. 393 pp.

Levine, Edwin Burton. *Hippocrates*. New York: Twayne Publishers, 1971.
Biography. Since little is known about the life of Hippocrates (c. 460-370 B.C.E.), the biographical portion of this book is short and aptly titled "The Mystery of Hippocrates and the Hippocratic Collections." In most of the text, Levine, a classics scholar, explains the content of the collection and its likely provenance, including the Hippocratic philosophy of medicine, how the ideal physician practices his profession, the management of diseases, and the environment of disease. End notes; bibliography; index. 172 pp.

Lipmann, Fritz. *Wanderings of a Biochemist*. New York: Wiley-Interscience, 1971.
Scientific autobiography. Fritz Lipmann confines his reminiscences to his studies in medicine and biochemistry and his career as a researcher in Germany, Denmark, and the United States, where he was at Rockefeller University. During his extremely productive career, he studied metabolic process, especially enzyme action and biosynthesis, which he explains for readers with training in biochemistry, and in 1953 shared the Nobel Prize for Physiology or Medicine with Hans Krebs. Reprints of his most significant articles fill about half of the book. Photos, drawings, and diagrams; end notes for each chapter. 229 pp.

McCormick, Joseph B., and Susan Fisher-Hoch, with Leslie Alan Horvitz. *Level 4: Virus Hunters of the CDC*. Atlanta, Ga.: Turner Publishing, 1996.
Memoirs. Long-time field researchers for the Centers for Disease Control, McCormick and Fisher-Hoch battled viral epidemics in South America, the United States, Africa, and Pakistan. Among the diseases they helped isolate were Lassa fever, AIDS, Crimean Congo Hemorrhagic Fever, and the dreaded Ebola. To handle these diseases they had to don protective suits and work in special laboratories—the "level 4"

protective condition—and the difficulty was compounded by poor conditions of field research. The authors spend much of the book revealing how devastated infrastructure and economies, political turmoil, and poor education created nearly insuperable obstacles in identifying and treating victims of these diseases. Prejudice also frustrated efforts, especially in research concerning the origin and spread of AIDS in Africa. An eye-opening book for all readers interested in epidemiology and public health. Photos; index. 379 pp.

Macfarlane, Gwyn. *Alexander Fleming, the Man and the Myth.* Cambridge, Mass.: Harvard University Press, 1984.
Biography. Alexander Fleming (1881-1955) is among the most honored medical researchers in the twentieth century. His discovery of penicillin brought him a share of the 1945 Nobel Prize for Physiology or Medicine, a knighthood, medals, and honorary degrees. Macfarlane insists most biographies of him, however, romanticize his work and personality and thereby obscure real mysteries about his story. In particular, she wonders, how exactly was penicillin discovered and why was it not developed during the following 12 years? Most of all, how did Fleming attract the fame when it was Howard Florey (also a 1945 Noble laureate) who proved its therapeutic value? She explores these questions as she lays out the family background, boyhood, education, and bacteriological work of Fleming. Her chapters about the discovery of penicillin show it to have come from an improbable chain of events, almost an accident. The rest of the book explains Fleming's public persona. This book is a first-rate example of investigative biography; it does not diminish Fleming so much as make him seem lifelike and rescue the story of penicillin from the shadows of legend. Photos and drawings; bibliography; end notes; index. 304 pp.

Mahone-Lonesome, Robyn. *Charles Drew.* New York: Chelsea House, 1990.
Biography for readers 12 years and older. The author describes the sudden need for blood to transfuse war wounded during World War II and the intense, successful efforts of Charles Drew (1904-1950) to meet that need. He

developed a method for separating plasma from blood, making long-term storage possible. After attracting attention as a high school and college (Amherst) athlete, Drew became an eminent surgeon, leader in establishing blood banks, and medical educator. Yet, as Mahone-Lonesome points out, the American Medical Association would not grant Drew a membership because he was African American. He did, however, win many awards and inspired many African American medical students before his death in a car accident. A movingly told story of hard work, brilliance, and perseverance. Photos; bibliography; index. 109 pp.

Marquardt, Martha. *Paul Ehrlich*. New York: Henry Schuman, 1951.

Biography. Best known for developing Salvarsan, the "magic bullet" that cured syphilis, Paul Ehrlich (1854-1915) conducted basic research in the theory of immunity and painstakingly determined the dosage levels of many medicaments. His work on immunity won him the 1908 Nobel Prize for Physiology or Medicine. Both celebrated and pilloried in his times, Ehrlich was, in Marquardt's skillful account, a meticulous researcher who shunned the limelight and worked hard because his natural optimism held that the conditions for human life can be improved. Readers with a knowledge of basic chemistry will profit most from the narrative, although much of the narrative is understandable to general readers. Photos. 255 pp.

Martin, Franklin H. *The Joy of Living*. 2 vols. Garden City, N.Y.: Doubleday, Doran, 1933.

Autobiography. Franklin H. Martin writes about his distinguished career in abdominal and gynecological surgery so that general readers will know better the problems encountered daily by him and his colleagues. He also wants to chronicle the change to scientific medicine and the birth of preventative medicine. He tells his life story in full, from his Wisconsin boyhood to his colonelcy in the Medical Corps during World War I, and along the way explains how he helped found the American College of Surgeons and his role in the surgical journal *Surgery, Gynecology, and Obstetrics*. Photos; index for each volume.

Mayo, Charles H. *Mayo: The Story of My Family and My Career.* Garden City, N.Y.: Doubleday, 1968.
Autobiography. Charles W. Mayo was the third-generation scion of a renowned medical family. His grandfather was a pioneer surgeon in the Midwest; his father, Charles H. Mayo, and his uncle, William J. Mayo, started the Mayo Clinic in Rochester, Minnesota. This book tells about them all, and their large family, and it is a stirring story of zestful go-getters, delightfully written and vivid in its portrayal of medical practices and research. Mayo also served as an ambassador on two occasions and knew a host of prominent politicians and scientists, if not from his practice at the clinic, then from his participation in politics. An absorbing saga for all readers. Photos. 351 pp.

Nolan, Jeannette Covert. *Florence Nightingale.* New York: Julian Messner, 1946.
Biography for readers 14 years and older. Written with novelistic touches, this book idolizes Florence Nightingale, who captured the admiration of England for her care of British soldiers during the Crimean War. She also improved the sanitation of hospitals, founded a nursing school, and championed the poor. Drawings. 209 pp.

Paul, Oglesby. *Take Heart: The Life and Prescription for Living of Dr. Paul Dudley White.* Boston, Mass.: Harvard University Press/Francis A. Countway Library of Medicine, 1986.
Biography. One of world's premier heart specialists, Paul Dudley White (1886-1973) emphasized preventive measures for a healthy heart and life: optimism, physical exercise, and productive work. Paul explains this simple but influential philosophy as he relates White's background and education, research and publications, professorship at Harvard University, and participation in such associations as the American Heart Association, which he helped found and served as president, and the World Congress of Cardiology. He had many famous patients, among them President Dwight D. Eisenhower. White also traveled extensively and became the first American member of the Soviet Union's Academy of Medical Science. For general readers Paul illustrates a nearly

ideal career of a subspecialist in mid-twentieth-century America. Photos; end notes; index. 315 pp.

Peters, C. J., and Mark Olshaker. *Virus Hunter: Thirty Years of Battling Hot Viruses Around the World.* New York: Doubleday, 1997.
Autobiography. With Olshaker's help, Peters tells the story of his career in the Army's Disease Assessment Division and at the Centers for Disease Control. Pursing deadly viruses, he traveled widely in Africa and South America from the 1960's into the 1990's. Among his preys were the feared Ebola and Machupo viruses. But he also played a key role in isolating the hantavirus that suddenly struck down Navahos in New Mexico in 1993. The stories are exciting in their own right and ably told in this book. The authors spice the narrative with much barracks-room vulgarity—sometimes, it seems, gratuitously so—but the style is punchy and vivid nevertheless. Readers will learn much about viruses but even more about the institutions that battle them and the national and international politics involved. Photos. 323 pp.

Plesset, Isabel R. *Noguchi and His Patrons.* Rutherford, N.J.: Fairleigh Dickinson University Press, 1980.
Biography. Plesset tells an unusual and dramatic story. Hideyo Noguchi (1876-1928) was the son of poor farmers in Japan and nearly lost his hand in a childhood accident, but hard work and driving ambition placed him among the most famous microbiologists of the era by the time he was 35 years old. He worked at the Rockefeller Institute on infectious diseases, such as yellow fever, of which he died during field research in Africa. His work created a controversy, and there were hints that he committed suicide or even was murdered because of it. Plesset carefully tells of Noguchi's family, boyhood, education, work as a physician in Japan and China, tenure at the Rockefeller Institute, and trips to Central and South America and West Africa. Largely unknown in America, Noguchi is a folk hero in Japan, according to Plesset, who reintroduces him to Western readers and clears up the misinformation about him that previous biographies contain. Photos; bibliography; end notes; index. 314 pp.

Prieto, Jorge. *Harvest of Hope*. Notre Dame, Ind.: University of Notre Dame Press, 1989.
Autobiography. Born in Mexico in 1918, Jorge Prieto came to the United States when his father became a political refugee. Prieto attended medical school and practiced family medicine at Cook County Hospital in Chicago, eventually directing the family practice department. He tells of dealing with his bicultural background and serving poor patients in a big city torn by social and racial strife. It is a hearteningly optimistic book for general readers. 157 pp.

Ptacek, Greg. *Champion for Children's Health: A Story About S. Josephine Baker*. Minneapolis, Minn.: Carolrhoda Books, 1994.
Biography for readers 12 years and older. Sarah Josephine Baker (1873-1945) was the first director of New York City's Division of Child Hygiene, itself the first government agency devoted to children's health. Credited with saving thousands of children's lives, she became a leading world expert on infant hygiene and introduced health care innovations, inaugurated child hygiene as a medical school topic, and lobbied for women's suffrage. Ptacek tells the story simply, emphasizing Baker's career. Drawings; bibliography. 64 pp.

Reid, Edith Gittings. *The Great Physician*. London: Oxford University Press, 1931.
Biography. A reverentially written account of the life of William Osler. To Reid, he was the acme of nineteenth-century doctors, the man who put medicine on a modern, scientific footing. She discusses his family, education, years as a professor at Canada's McGill University, family life, professorships at Johns Hopkins and Oxford, knighthood, and views concerning World War I. Her portrayal makes him look both sternly progressive in his reforms of medical education and humanitarian in his disgust with war fervor and in his constant dedication to hospital patients from all levels of society. Index. 299 pp.

Robbins, Christine Chapman. *David Hosack, Citizen of New York*. Philadelphia, Pa.: American Philosophical Society, 1964.
Biography. David Hosack (1769-1835) was a prosperous

physician in New York City who built the Elgin Botanical Garden for research into the medicinal properties of plants. Not much came from the gardens, which lasted only ten years before the area was razed to make room for the spreading city. But Hosack's influence, according to Robbins, was great among prominent scientists and citizens of the day, and he also started his own medical school. This book gives history buffs a portrait of American medicine in the early nineteenth century as much as it profiles Hosack. Portraits and drawings; bibliography; index. 246 pp.

Salber, Eva J. *The Mind Is Not the Heart*. Durham, N.C.: Duke University Press, 1989.
Autobiographical essays. This volume collects 17 essays by Eva J. Salber (b. 1916), arranged in three sections. The first section concerns her medical education and practice among the poor in South Africa. The second section deals with her marriage and move to Boston, where she worked in public health. The final section describes her tenure as professor of community and family health at Duke University, where she studied the rural poor in North Carolina and their health problems. Photos; end notes. 282 pp.

Thomas, Lewis. *The Youngest Science: Notes of a Medicine-Watcher*. New York: Viking Press, 1983.
Autobiography. Few writers equal Lewis Thomas' graceful, amicable prose style, and this book is a pleasure to read throughout. As well as any autobiography, it brilliantly portrays the process of research, with all its excitement, wrong turns, disappointments, and sudden realizations. With affection and humor, Thomas (1913-1994) recalls his boyhood, father's medical practice, mother's nursing, medical education, military service as an epidemiologist, professorships, tenure as dean at New York University and Yale Medical School, and directorship of the Sloan-Kettering Cancer Center. Along the way, he tells of his frustration in dealing with New York politics, the wonders of microorganisms, and the sheer delight of having a difficult problem to solve. His descriptions of patients and their behavior in hospitals are unsurpassedly evocative. For all readers. End notes. 270 pp.

Thomson, Elizabeth H. *Harvey Cushing: Surgeon, Author, Artist.* New York: Henry Schuman, 1950.

Biography. In addition to being an innovative surgeon, Harvey Cushing (1869-1939) wrote an erudite biography of his friend and colleague at John Hopkins Medical School, William Osler. Cushing believed that doctors must do more than treat disease; they must understand the patient, and to do so they must be people of broad education and interests. Cushing illustrated that credo in his writing, cartoons and anatomical drawings, teaching, and research as well as in his surgical practice, as Thomas shows in this book for general readers. Photos and drawings; bibliography; index. 347 pp.

Waksman, Selman A. *My Life with the Microbes.* New York: Simon and Schuster, 1954.

Autobiography. Selman A. Waksman (1888-1973) discovered streptomycin in 1943. The powerful antibiotic helped modern medicine redirect its focus from palliating symptoms to curing disease, and for his contribution Waksman received the 1952 Nobel Prize for Physiology or Medicine. Here he recalls his life and career for general readers. The story begins in Russia, builds through his immigration to the United States in 1910 and education at Rutgers University, continues with his professorship there and work in bacteriology, and climaxes with his many awards and honors. Photo of Waksman. 359 pp.

Whitteridge, Gweneth. *William Harvey and the Circulation of the Blood.* London: MacDonald, 1971.

Scientific biography. In this close analysis, the author traces the development of William Harvey's scientific research and its culmination in the publication of *De Motu Cordis* (*On the Circulation of Blood,* 1627). The text begins with Harvey's medical training at Cambridge University and the University of Padua, emphasizing the influence of his teachers, and then covers his London medical practice, lectures on anatomy, the writing of his magnum opus, its reception, and the publication of *De Generatione Animalium* (*On the Generation of Animals,* 1651). Whitteridge writes primarily for other scholars. Drawings; bibliography; end notes; index. 269 pp.

Wilson, Dorothy Clarke. *I Will Be a Doctor: The Story of America's First Woman Physician.* Nashville, Tenn.: Abingdon Press, 1983.
Biography for readers ten years and older. Wilson tells the story of Elizabeth Blackwell (1821-1910) in the manner of a novel. It is an inspiring story. Something of a misfit, English-born Blackwell moved with her family to America. She resolved to become a physician even though such a career was unheard of for women at the time. She returned to England for education and then opened a clinic in New York City. Eventually, overcoming the disapproval of the medical establishment, she started the first medical school for women and was instrumental in improving sanitation in the city and hygiene in medical practices. 160 pp.

Wolfe, Rinna Evelyn. *Charles Richard Drew, M.D.* New York: Franklin Watts, 1991.
Biography for readers eight years and older. Wolfe describes how Charles Drew overcame biases about African Americans to become a world-famous physician, especially noted for his development of techniques to separate plasma from blood. During World War II this breakthrough in preserving blood saved the lives of many wounded. Wolfe emphasizes Drew's contributions to the education of African American doctors and the tragedy of his early death in a car accident. Photos and drawings; well-formulated glossary; bibliography; index. 64 pp.

Wyeth, John Allan. *With Sabre and Scalpel.* New York: Harper and Brothers, 1914.
Autobiography. Wyeth divides his book into two major sections. The first is a paean to the pre-Civil War South and recounts his war experiences as a Confederate officer. The second concerns the clinic he founded in New York City and his research in surgical treatment of tumors, skin transplantation, and joint maladies. Picturesquely written, the text is frank in conveying Wyeth's political and medical views. A vivid account of early American surgery. Photos and drawings. 535 pp.

SERIES FOR YOUNG READERS

Pioneers in Health and Medicine. New York: Twenty-First Century Books, 1991-1993.

Biographies of famous physicians and medical researchers for readers ten and older. Simply and clearly written, the books of this series often feature minority or women physicians: *The Life of Elizabeth Blackwell* by Elizabeth Schleichert (1992); *The Life of Dorothea Dix* by Schleichert (1992); *The Life of Charles Drew* by Katherine S. Talmadge (1991); *The Life of Alexander Fleming* by Judith Kaye (1993); *The Life of Louis Pasteur* by Marcia Newfield (1992); *The Life of Florence Sabin* by Kaye (1993); and *The Life of Benjamin Spock* by Kaye (1993). Photos; short bibliographies; indexes. About 80 pp.

Chapter 8

Physics

COLLECTIONS

Beyerchen, Alan D. *Scientists Under Hitler.* New Haven, Conn.: Yale University Press, 1977.
Biography-history. Although primarily a historical study, this book contains much valuable biographical information about German physicists, the study's focus, during the 1930's, information drawn primarily from unpublished documents and interviews. Beyerchen's purpose is to show how physicists responded to the political turmoil caused by the educational and racial policies of the Nazis. It is a frightening story, and among the personalities are some of the greatest physicists of the twentieth century, such as Max Planck, Max Born, Albert Einstein, and Werner Heisenberg, as well as lesser-known figures who once held deadly political power, such as Philipp Lenard and Johannes Stark. A scholarly history, the book is nevertheless illuminating for general readers. Photos; bibliography; end notes; index. 287 pp.

The Biographical Dictionary of Scientists: Physicists. New York: Peter Bedrick Books, 1984.
A historical sketch of physics prefaces alphabetized, short articles on 199 physicists, as early as Democritus (ca. 460-370 B.C.E.). Averaging about 1,000 words apiece, the articles offer basic biographical information and a summary of each scientist's achievements for general readers. A useful resource because of the extent of the coverage and the brevity of the articles. Drawings; glossary; index. 212 pp.

Boorse, Henry A., Lloyd Motz, and Jefferson Hane Weaver. *The Atomic Scientists: A Biographical History.* New York: John Wiley and Sons, 1989.

Starting with the Roman poet Lucretius (99-55 B.C.E.) and concluding with Robert Hofstadter (b. 1915), the authors construct the history of atomic physics through dozens of biographical sketches of its leading investigators. Concise but well-formulated and informative, the sketches cover the careers and personal lives of scientists and explain their contributions for a general adult audience interested in the history of science. The authors want to show that scientists, far from being dull and abstract people, find excitement in their work and have fascinating personalities because of it. Photos; short bibliographies at chapter ends; index. 472 pp.

Brennan, Richard P. *Heisenberg Probably Slept Here: The Lives, Times, and Ideas of the Great Physicists of the 20th Century.* New York: John Wiley and Sons, 1997.

Because of the simplicity of its style, the low level of technicality in its explanation of scientific ideas, and its focus on character eccentricities, this book is well suited to readers, high school students and older, who like spicy reading. In addition to an introduction that outlines the history of physics from its beginnings and a thoughtful epilogue about philosophical extensions of physical ideas, Brennan discusses Isaac Newton, Albert Einstein, Max Planck, Ernest Rutherford, Niels Bohr, Werner Heisenberg, Richard Feynman, and Murray Gell-Mann. Drawings; bibliography; glossary; index. 274 pp.

Crease, Robert P., and Charles C. Mann. *The Second Creation.* New York: Macmillan, 1986.

Biography-history. The book is a thorough, highly readable job of science journalism, often thrilling in its exposition of the development of atomic, nuclear, and particle physics. The authors based their story on extensive interviews and research. The interviews thus provide much biographical detail not found in previously published volumes. Dozens of physicists make an appearance—from Joseph John Thomson to Stephen Weinberg. Written for a general audience, the book excels at revealing complex scientific theories and discoveries lucidly but in non-technical language; it also chronicles the politics and battles of personalities that lie behind some scientific experiments and the interpretations of their data. Diagrams; end notes; glossary; index. 480 pp.

Heathcote, Niels H. de V. *Nobel Prize Winners in Physics 1901-1950*. New York: Henry Schuman, 1953.
The chronologically ordered essays start with Wilhelm Conrad Roentgen in 1901 and end with Cecil Frank Powell in 1950. Because each essay is devoted to one award year, cowinners for that year are grouped together, and the biographical information about them tends to be sketchier than for essays about unshared prizes. Each essay provides a biographical sketch, a description of the prize-winning discoveries, and the practical and theoretical consequences of the discoveries. The scientific explanations, often quoted from other writers, are moderately technical, sometimes containing chemical notation or sophisticated mathematics. Photos and diagrams; index. 473 pp.

Nobel Prize Winners: Physics. 3 vols. Pasadena, Calif.: Salem Press, 1989.
The articles in this collection cover the 133 laureates from 1901 to 1988. An introductory essay explains the history and administration of the Nobel Prize, and a handy time line lists the winners in chronological order. The articles, also in chronological order, run about 3,500 words. Each contains a list of vital statistics and summary of the laureate's prize-winning contribution to physics; short sections of the article recount the occasion of the award, the laureate's Nobel lecture, and major events of his or her personal life; a substantial section concerns the scientific career. Each essay concludes with a bibliography of the laureate's principal publications and an annotated bibliography of biographical sources. Photos; index in vol. 3.

Schweber, Silvan S. *QED and the Men Who Made It: Dyson, Feynman, Schwinger, and Tomonaga*. Princeton, N.J.: Princeton University Press, 1994.
Written for readers who understand advanced physics and

mathematics, this book explains the development of quantum electrodynamics (QED), often called the most successful theory in physics. Schweber provides historical background as well as sophisticated explanations of the physical principles. In separate sections, he also provides analytical sketches of the careers of Richard Feynman (1918-1991), Julian Schwinger (b. 1918), Sin-itiro Tomonaga (1906-1979)—who shared the 1965 Nobel Prize for Physics—and Freeman Dyson (b. 1923). Photos and diagrams; extensive bibliography; end notes; index. 732 pp.

Serafini, Anthony. *Legends in Their Own Time: A Century of American Physical Scientists*. New York: Plenum Press, 1993.
Biography-history. Serafini wants to show that science is more than an accumulation of facts and theory, that it involves human personalities and drama. Accordingly, he opens his account of physics, chemistry, and astronomy in America (often featuring foreign-born scientists) in the 1830's at Harvard Observatory and moves forward with the lives and achievements of leading minds in each generation, ending with a chapter about work on grand unification theories (GUTs). Among the scientists included are Josiah Gibbs, Albert Michelson, Robert Millikan, Linus Pauling, Han Bethe, Murray Gell-Mann, and Richard Feynman. Serafini makes no claims to be comprehensive and is not, but the text is casually and entertainingly written, if fragmented, and he explains the science for general readers who do not want to be immersed in details. End notes; index. 361 pp.

Snow, C. P. *The Physicists*. Boston: Little, Brown, 1981.
Biography-history. This book is primarily a history of physics in the twentieth century, but along the way there is much biographical information about prominent physicists. A scientist, administrator, and philosopher of science, Snow knew many of the people he writes about. Among the subjects are Marie and Pierre Curie, Joseph John Thomson, Ernest Rutherford, Peter Kapitsa, Albert Einstein, Niels Bohr, Paul A. M. Dirac, Werner Heisenberg, Max Born, Enrico Fermi, Lise Meitner, J. Robert Oppenheimer, Edward Teller, Richard Feynman, and Abdus Salam. The biographical information largely comprises

anecdotes and judgments about scientific abilities. Snow died before he could revise and polish the manuscript. It often seems desultory and overburdened with his concerns about the promise and danger of modern technology. Photos; index. 192 pp.

Weber, Robert L. *Pioneers of Science.* Bristol, England: Adam Hilger, 1988.
Weber provides short biographical sketches of Nobel laureates in physics from the first prize in 1901 until 1987. The sketches concentrate on the prize-winning discoveries. He also summarizes the career of Alfred Nobel and the bequest for his prizes. Portraits of each Nobelist; bibliography; index. 310 pp.

BIOGRAPHIES AND AUTOBIOGRAPHIES

Abragan, Anatole. *Time Reversal.* Oxford, England: Clarendon Press, 1989.
Autobiography. Written with considerable charm and wit, this book is also often not easy reading for non-scientists. Anatole Abragan (b. 1914) was a theoretical physics professor at the Collège de France and later director of physics at the Commissariat à l'Energie Atomique, and his discussions of atomic physics, nuclear magnetism, bosons and resonances, and accelerators are sophisticated, although not purely technical. He also relates his childhood in Russia and France, education, and travels in the service of physics to England, Russia, India, Asia, and America. Photos; index. 373 pp.

Ajzenberg-Selove, Fay. *A Matter of Choices: Memoirs of a Female Physicist.* New Brunswick, N.J.: Rutgers University Press, 1994.
Autobiography. Born in 1926 in Poland and raised in France, Ajzenberg-Selove had to flee with her family to the United States to escape the Nazi occupation. Despite indifferent grades in high school and college, she became a highly respected experimental nuclear physicist, teaching at Haverford College and the University of Pennsylvania. With

considerable graciousness and candor, she tells the story of her quest to find her way in a new country and enter a male-dominated profession. The chapters about her childhood, education, and marriage to Walter Selove are especially engaging; the rest of the book is less cohesive, although readers will find unusual perspectives on such famous physicists as Edward Teller, Victor Weisskopf, and William Fowler. Ajzenberg-Selove does not dwell on the particulars of her scientific contributions, describing them for general readers, but focuses on her relations to other researchers, pioneering teaching career, and the politics of physics. Photos; end notes; index. 234 pp.

Alvarez, Luis W. *Alvarez: Adventures of a Physicist.* New York: Basic Books, 1987.
Autobiography. Nobel laureate Luis Alvarez (1911-1988) shows little introspective talent in this book; it is primarily a parade of his scientific accomplishments, inventions, friendships, and government work. It is an incredible parade. Alvarez worked on the first big science project, the cyclotron, with Ernest O. Lawrence, his mentor. He contributed to the development of radar and the atomic bomb (as a member of the Manhattan Project). He invented the hydrogen bubble chamber, which was crucial to particle physics. He punctured several elements of conspiracy theories about the assassination of President John F. Kennedy. With his son, Walter, he developed the asteroid impact theory to explain the extinction of dinosaurs. He knew nearly all the important physicists of the mid-twentieth century. For these reasons the book, although sometimes woodenly written, is fascinating, an invaluable look at the growth of modern physics by one of its best experimentalists. It is suitable for general readers, although a knowledge of basic physics will help because the extensive science explanations are sometimes technical. Photos; index. 292 pp.

Arms, Nancy. *A Prophet in Two Countries.* Oxford, England: Pergamon Press, 1966.
Biography. Arms offers general readers a close look at Franz E. Simon (1893-1956). Although a hero in World War I, Simon

had to flee his native Germany because of the Nazi persecution of Jews before World War II and found sanctuary in England, where he worked at the Clarendon Laboratory at Oxford. During the war he participated in the gas diffusion project to purify weapon-grade uranium for the Manhattan Project. Arms recounts these events and Simon's background in Berlin, education, and research in low-temperature physics. Photos; bibliography; index. 171 pp.

Bernstein, Jeremy. *Hans Bethe, Prophet of Energy*. New York: Basic Books, 1980.
Biography. One of the founders of quantum physics and of astrophysics, Bethe was the leading theoretician in the Manhattan Project and a leader in post-World War II development of atomic energy. Based on extensive interviews with Bethe and others, Bernstein gives educated general readers an entertaining, sophisticated portrait of the man, his ideas, and the scientific and social context of those ideas. The biography falls into three parts, the first concerning quantum mechanics, the second the Manhattan Project and hydrogen bomb, and the third atomic energy and policy regarding it. Bibliography; index. 212 pp.

Biquard, Pierre. *Frédéric Joliot-Curie: The Man and His Theories*. New York: Paul S. Eriksson, 1965.
Biography. Biquard extols the scientific discoveries and political activity of his friend, Frédéric Joliot-Curie, who shared the 1935 Nobel Prize for Chemistry with his wife, Irène, for producing the first artificial radioactivity. Joliot-Curie also performed experiments that led to the discovery of the neutron and atomic fission; after World War II he helped develop atomic energy for his homeland, France. During the war, he participated in the French anti-occupation movement and joined the French communist party. Biquard's account, which suffers from the stiffness of an inelegant translation, treats Joliot-Curie's childhood, marriage and scientific career, political activities, and personality in separate chapters. A large selection of Joliot-Curie's writings conclude the book. Short bibliography; index. 192 pp.

Bitter, Francis. *Magnets: The Education of a Physicist*. New York: Doubleday, 1959.
Autobiography-popular science. A curious and charming idea for a book: Francis Bitter offers to teach non-scientists the principles of magnetism by recounting his education at Columbia University and life as a physicist. There is more about magnetism and atomic physics than about how he made it through his exams and dissertation, but the text is humorous and casual and conveys much information pleasantly. Drawings; index. 155pp.

Blaedel, Niels. *Harmony and Unity*. Madison, Wis.: Science Tech Publishers, 1988.
Biography. Translated from the Danish edition, this book is a portrait of Niels Bohr (1885-1962) for non-physicists by a fellow Dane, who was able to interview many of Bohr's friends and family. For explaining the electron structure in the atom, Bohr received the 1922 Nobel Prize for Physics, and then he helped develop the standard interpretation of quantum mechanics and extended his key concept, complementarity, into general philosophy. Such contributions left a deep imprint on the history of physics and the philosophy of science, especially since Bohr's institute in Copenhagen served as a sort of headquarters for physics in the first half of the twentieth century. Blaedel recounts all these developments after a close look at Bohr's boyhood, family, education, extensive travels, and relations with the other great physicists of the era, especially Albert Einstein and Werner Heisenberg. He also discusses Bohr's work for world peace. Photos and drawings; bibliography; index. 323 pp.

Blumberg, Stanley A., and Louis G. Panos. *Edward Teller: Giant of the Golden Age of Physics*. New York: Charles Scribner's Sons, 1990.
Biography. Praised for his extraordinary brilliance all around, denounced as a warmonger by anti-nuclear activitists, and shunned by many of his contemporaries for his testimony during a hearing on J. Robert Oppenheimer's security clearance, Edward Teller (b. 1908) has been one of the most controversial scientists of the century. Based upon extensive inter-

views with Teller, physicists, and government officials, the authors dispel much of the misinformation about Teller, especially concerning his roles in the Manhattan Project, development of the hydrogen bomb, and Oppenheimer hearing. In doing so, they show him to be a tireless fighter, scientifically and politically, for his views, especially his support for President Ronald Reagan's Strategic Defense Initiative. In fact, the book is as much a journalistic tour of United States strategic defense policies as a biography of Teller. Photos; bibliography; end notes; index. 306 pp.

Bonner, Elena. *Alone Together.* New York: Knopf, 1986.
Memoirs. Bonner, a physician, was the wife of Andrei Sakharov, who had the unusual distinction of being the father of the Soviet hydrogen bomb and of winning a Nobel Peace Prize. The book focuses on their internal exile in the Soviet Union, imposed for protesting government policies, especially the treatment of Jews and the stockpiling of atomic weaponry. There is also information about their families and upbringing. Unfortunately, the text is confusing and will best be appreciated by readers already familiar with the life of Bonner and Sakharov. Photos; index. 270 pp.

Born, Max. *My Life: Recollections of a Nobel Laureate.* New York: Charles Scribner's Sons, 1978.
Autobiography. One of the great theoretical physicists of the twentieth century, Born (1882-1970) is especially honored for his contributions to the mathematical structure of quantum mechanics. These memoirs, originally written for his family, recount in considerable detail his youth and education in Germany, university career, relationships with such colleagues as Albert Einstein and David Hilbert, treatment by the Nazis, and flight to England before World War II. His account of quantum mechanics assumes a physicist's level of knowledge and is likely to seem mostly a series of odd terms and equations to general readers. Still, all readers will learn much about the society and politics of physical researchers during the first half of the century, because Born knew nearly every leading physicist and collaborated with many. The tone is formal and courteous, and the style a little labored in Born's

concern for balance and accuracy, so that the book sounds very much like a European professor of a bygone era. Photos; index. 308 pp.

Brian, Denis. *Einstein: A Life.* New York: John Wiley and Sons, 1996.
Biography. Brian draws from fresh interviews with friends and colleagues of Albert Einstein and from material in the Einstein Archives released during the 1980's to bring a new perspective to many long-standing questions about him. Brian especially argues that Einstein was not politically naive, as many biographers claim, and was not coldhearted to his family. Brian confirms but plays down Einstein's reputation as a womanizer and examines the evidence concerning his alleged illegitimate children. The book does not discuss physics in depth, concentrating instead on Einstein's private life and those aspects of his public image which Brian feels have been distorted. As such, this book is a useful supplement to earlier biographies, such as those by Abraham Pais and Ronald William Clark (see below). Photos; end notes; bibliograhy; index. 509 pp.

Brinitzer, Carl. *A Reasonable Rebel: Georg Christoph Lichtenberg.* New York: Macmillan, 1960.
Biography. Georg Christoph Lichtenberg (1742-1799) overcame a deadly childhood illness which deformed his spine and became a professor of astronomy, mathematics, and physics at the University of Göttingen. Brinitzer tells Lichtenberg's story with consistent charm and drama for all readers, but the scientific career is less compelling than Lichtenberg's vitality and breadth of interests. He was a controversialist, witty writer, and experimentalist—not only with physical apparatus but also with dreams. He was also an unusually frank diarist, and the diaries enable Brinitzer to depict the racy side of eighteenth-century society, both in Germany and in England, which Lichtenberg visited. Bibliography; index. 203 pp.

Broda, Englebert. *Ludwig Boltzmann, Man, Physicist, Philosopher.* Woodbridge, Conn.: Ox Bow Press, 1983.

Biography. First published in German in 1955, this book is a sophisticated explanation of the physical and philosophical ideas of Ludwig Boltzmann (1844-1906). The author opens with a synopsis of Boltzmann's life and beliefs, then addresses his work in thermodynamics, statistical mechanics, and atomism. Pondering the nature of statistical mechanics, Boltzmann realized that our understanding of nature is essentially hypothetical, and the author examines the ramifications of Boltzmann's philosophical insights in the book's last sections. For educated readers, especially those with an interest in intellectual history. Photos; bibliography; end notes; indexes. 169 pp.

Brown, Sanborn C. *Benjamin Thompson, Count Rumford.* Cambridge, Mass.: MIT Press, 1979.
Biography. Is there a stranger story among the biographies of scientists than that of Benjamin Thompson (1753-1814)? From a Massachusetts farmer's family, young Thompson started early to experiment and educate himself in science, and he was brilliant at it. But he was also ridden by ambition, a shameless self-aggrandizer, and ruthless—often a liar when it benefited him. During the American Revolution he was a loyalist spy, became a soldier, then in England became a bureaucrat, rose to the rank of colonel, was knighted reluctantly by George III, moved to Bavaria, where he reformed the army and set up a system of poor houses that put beggars to work and was named a count of the Holy Roman Empire (Count Rumford), moved back to England and started the Royal Institution, and finally settled in France during the Napoleonic wars. He was haughty, overbearing, pompous, and considered a friend of the poor. Along the way he had many mistresses, two wives whom he abandoned, including the widow of Antoine Lavoisier, and troubles over various scandals involving malfeasance or plagiarism. He vastly improved kitchen stoves and fireplaces, lighting fixtures, and thermometers of all kinds; his investigations into the propagation of heat led to the modern theory of radiant energy. Even the count's death has a tang of mystery and intrigue. He was a remarkable man obsessed with reputation, science, and wealth, and Brown tells the story with delicacy and insight,

explaining the count's scientific ideas with consistent clarity. Photos, drawings, and diagrams; end notes; index. 361 pp.

Bucky, Peter A., with Allen G. Weakland. *The Private Albert Einstein*. Kansas City, Mo.: Andrews and McMeel, 1992.
Biography. Peter Bucky's father, Gustav Bucky, was among Albert Einstein's oldest and closest friends; the Einsteins and Buckys often spent time together. The author thus has an unusual perspective on Einstein and a good deal of firsthand information, which he publishes in this "intimate sketch." Sections of narrative and question-and-answer style conversations alternate in chapters devoted to Einstein the man, Einstein in America, his views on Germany, the atomic bomb project, his ideas about religion and education, his family, and his music and comic verse. For readers already familiar with Einstein's life, this is an enjoyable supplement. Photos; index. 171 pp.

Cantor, Geoffrey. *Michael Faraday: Sandemanian and Scientist*. New York: St. Martin's Press, 1991.
Biography. Cantor scrutinizes the beliefs of the Sandemanian Protestant sect in order to connect those beliefs to the life and ideas of its most famous member, Michael Faraday (1791-1869). He concludes that Faraday's religious beliefs formed the foundation for his scientific theories, including his work on electricity and magnetism. Thus, much of the book involves theology, but Faraday's family background, experiments, and approach to science receive attention as well. Photos and diagrams; bibliographic essay; end notes; index. 359 pp.

Cardwell, Donald S. L. *James Joule*. Manchester, England: Manchester University Press, 1989.
Biography. Cardwell considers James Joule (1818-1889) a bridge figure between the time when science was a gentleman's avocation and the time it became a profession. Cardwell shows that Joule was a man of diverse talents in science, although he is now known for discovering the conservation of energy and for the energy unit named after him. He explains Joule's work in depth, which included research in

meteorology and electricity. The book addresses his youth in Manchester, education, presidency of the British Association, and dealings with such scientists as Heinrich Helmholtz and William Thomson (Lord Kelvin). Readers should understand basic physics for this book. Photos and diagrams; end notes; indexes. 333 pp.

Caroe, G. M. *William Henry Bragg, 1862-1942: Man and Scientist.* Cambridge, England: Cambridge University Press, 1978.
Biography. What sort of a person is a scientist, what leads to discovery, and how is responsibility taken for the discovery? Caroe pursues such questions as she tells the story of her father, William Henry Bragg. With his son, William Lawrence Bragg (who helped plan this book), Bragg won the 1915 Nobel Prize for Physics for founding X-ray crystallography, a fundamental tool in exploring the structure of matter. She discusses his youth and education, professorships in Australia and England, tenure at the Royal Institution, work during World War I, and contributions to education and industry as well as his investigations of crystals. Photos; end notes; index. 212 pp.

Casimir, Hendrik B. G. *Haphazard Reality.* New York: Harper and Row, 1983.
Autobiography. A series of anecdotes primarily, this book also contains short essays on developments in modern physics, technology, and education from Casimir's point of view. It is an intimate point of view, because he participated in the development of quantum mechanics in the late 1920's and 1930's, and he writes in a modest, cultivated tone. For the most part his subjects are the scientists he knew and worked with, many of them now legendary, such as Niels Bohr, Paul Ehrenfest, and Wolfgang Pauli. After World War II, Casimir worked as an industrial researcher and administrator, and he discusses the relation of science and industry. Bibliographical notes; index. 356 pp.

Cassidy, David C. *Uncertainty: The Life and Times of Werner Heisenberg.* New York: W. H. Freeman, 1992.
Biography. A well-titled book. Heisenberg (1901-1976) had

one of the most successful and troubling careers in physics. He formulated the famous uncertainty principle, co-authored quantum mechanics, and with Danish physicist Neils Bohr established the prevailing interpretation of quantum mechanics—towering achievements. He also helped lead the German effort to develop atomic power and weapons during World War II, and the exact nature of his compromises with the Nazi government remains controversial. Cassidy treats his subject with balance but also with some skepticism about Heisenberg's own defense of his wartime activities. He shows Heisenberg to be an idealist in physics, an overachiever in his youth to impress a demanding father, and a German patriot, although no Nazi enthusiast. Heisenberg's achievements in science receive careful explication, but the book's focus rests on the cultural atmosphere that shaped Heisenberg. Photos; end notes; indexes. 669 pp.

Childs, Herbert. *An American Genius: The Life of Ernest Orlando Lawrence*. New York: E. P. Dutton, 1968.
Biography. Ernest O. Lawrence (1901-1958) invented the first big atom-smasher, the cyclotron, which brought him celebrity and the Nobel Prize for Physics while he was still a young man. He assembled a radiation laboratory at the University of California, Berkeley, and trained there another generation of brilliant scientists. He was a key figure in the Manhattan Project and a political foe of J. Robert Oppenheimer afterwards. Clearly, as Childs argues, he is among the most important American scientists of the twentieth century. Childs presents Lawrence as a distinctively American phenomenon: from immigrant stock, raised in a small town in South Dakota, educated at Yale University, and intent on the defense of his country. Childs also emphasizes Lawrence's vitality and enthusiasm for physics, which he elucidates non-technically. For general readers. Photos; bibliography; index. 576 pp.

Clark, Ronald William. *Einstein: The Life and Times*. New York: World Publishing, 1971.
Biography. A skillfully written study of Einstein's entire life (1879-1955). Clark devotes about equal time to explanations of Einstein's scientific theories, politics, and personal life.

Extensive and judicious use of source material produces a balanced account, as Clark takes care to correct legends and discountenance apocryphal stories. Einstein's hearty sense of humor comes out clearly in the narrative, and thanks to the clarity of Clark's summaries, so do the major ideas in relativity and quantum theory. Einstein's participation in the Zionist movement and his battle with the Nazi government of Germany receive especially full treatment. Photos; end notes; bibliographic essay; index. 718 pp.

Cohen, Bernard I. *Benjamin Franklin's Science*. Cambridge, Mass.: Harvard University Press, 1990.
Scientific biography. As Cohen points out, as famous as Benjamin Franklin is, there are few books that thoroughly explain his scientific interests, much celebrated during his era. Cohen rectifies this lack by explaining the origins of Franklin's interests; his style of research and reasoning; the nature of his experiments with electricity, lightning, and heat; his invention of the Franklin stove and the lightning rod; and his observation of the transit of Mercury. The book is a readable, valuable addition to Franklin studies and to the history of science in America. Drawings and diagrams; end notes; index. 273 pp.

Coulson, Thomas. *Joseph Henry, His Life and Work*. Princeton, N.J.: Princeton University Press, 1950.
Biography. Joseph Henry (1797-1878) discovered electromagnetic induction and self-induction, had a fundamental unit named after him, and greatly improved the power and utility of electromagnets. He was the foremost physical scientist in America during the mid-nineteenth century. Yet, as Coulson points out, Henry is an obscure figure compared with his British counterpart in science, Michael Faraday. America was a scientific backwater at the time, and Henry, the son of a laborer, had to struggle to educate himself. He succeeded, and Coulson unfolds the unusual story of that success—Henry's career as a scientist occurred at a time that saw very few such careers in America. He helped found the National Academy of Sciences and Smithsonian Institution and worked on telegraphy. Coulson's account depicts the practice and use of

science in the nineteenth century through Henry's career. Photos; bibliography; index. 352 pp.

Crawford, Deborah. *Lise Meitner, Atomic Pioneer.* New York: Crown Publishers, 1969.
Biography for readers 14 years and older. Crawford opens her story of Lise Meitner in 1900, when, 22 years old, Meitner decided to attend the University of Vienna to become a physicist. The following 18 chapters relate her life for periods of one to five years, during which she performed crucial experiments with Otto Hahn and others that proved the nucleus of the atom could be split. She then had to flee Nazi persecution before she could complete her work. The last chapter takes readers from World War II until her death in 1968 and the belated recognition of her contribution to modern physics. Bibliography; index. 192 pp.

Curie, Marie. *Pierre Curie.* New York: Macmillan, 1923; Dover, 1963.
Biography-autobiography. Marie Curie recalls the life of her husband and scientific partner, Pierre Curie (1859-1906), but the last one-third of the book contains purely autobiographical notes. Drawing from her knowledge of him and letters and occasional writings, she delivers a portrait that is an unusual mixture of intimacy and objectivity. Certainly, she was qualified, as no one else, to judge his scientific ability; additionally, she supports her high opinion with quotations from other French scientists. The best passages of this short account describe their work together on radium and other radioactive materials. The autobiographical notes are modest in tone and concentrate on her education and scientific achievements. Photos. 118 pp.

Dank, Milton. *Albert Einstein.* New York: Franklin Watts, 1983.
Biography for readers 12 years and older. To present the complex life of Albert Einstein clearly for young readers, Dank devotes chapters to Einstein's youth, education and the classical physics taught at the time, relativity, professorships in Switzerland and Germany, increasing fame and troubles living in Nazi Germany, the revolution he started in physics,

disagreement with other physicists about quantum mechanics, and years in the United States. Dank writes effectively about the basic principles of Einstein's theories. Photos and diagrams; bibliography; index. 122 pp.

Davis, Nuel Pharr. *Lawrence and Oppenheimer*. New York: Simon and Schuster, 1968.
Biography-history. This book studies the contrasts, and eventual conflict, between Ernest O. Lawrence and J. Robert Oppenheimer. Lawrence was a conservative westerner and experimental physicist who designed early atom smashers; Oppenheimer was a liberal easterner and theoretical physicist who became an administrator. They collaborated in setting up and running the Manhattan Project during World War II but eventually disagreed over security and the fate of the project after the war. Lawrence was involved in the hearings during which Oppenheimer was accused of treason and lost his security clearance. Not a glorious passage in American history or science, the Lawrence-Oppenheimer affair exemplifies the suddenly brittle politics of physics during the Cold War. Davis summarizes Lawrence's career in the first third of the book and then focuses on the atomic bomb project and its aftermath. Bibliography; glossary; end notes; index. 384 pp.

De Haas-Lorentz, G. L., ed. *H. A. Lorentz: Impressions of His Life and Work*. Amsterdam, The Netherlands: North-Holland Publishing, 1957.
Biographical essays. If ever there was saint in physics, it was Hendrik Antoon Lorentz (1853-1928), winner of the second Nobel Prize for Physics (with Pieter Zeeman) and a principal transition figure between classical and modern physics. He was a theorist and took his place among the first rank of thinkers by extending James Clerk Maxwell's theory of electromagnetism to empty space, explained the splitting of lines of a spectrum by a magnetic field, and formulated the Lorentz-Fitzgerald contraction concerning light. These essays emphasize the clarity of his mind and kindness, gentlemanly demeanor, serenity, leadership, and breadth of knowledge. The first, by Albert Einstein, reveres him as a mentor and credits his work in electromagnetism as a necessary step

toward the special theory of relativity. Other essays, by colleagues and family members, outline the principal events in his life. That there is no full biography of Lorentz in English for a general readership is a major gap in scientific biography. Photos and drawings. 172 pp.

Dommermuth-Costa, Carol. *Nikola Tesla: A Spark of Genius.* Minneapolis, Minn.: Lerner Publications, 1994.
Biography for readers 12 years and older. Nikola Tesla (1856-1943) discovered how to induce a rotating magnetic field, the principle that led to his most famous invention, the alternating-current motor. The author presents his story—including his eccentricity and reclusiveness—with dramatic flair but is also careful to explain basic electricity for young readers. Photos and drawings; bibliography; index. 143 pp.

Dorozynski, Alexander. *The Man They Wouldn't Let Die.* New York: Macmillan, 1965.
Biography. A journalistic portrait of Russia's most versatile and penetrating theoretical physicist in the twentieth century, Lev Davidovich Landau (1908-1968). For the most part the author covers Landau's life and achievements superficially and even-handedly, although he devotes an entire chapter to the Soviet Union's development of nuclear weapons. The book concludes with a long account of Landau's near-fatal car accident and the "miracle of Moscow" operation that extended his life; unfortunately, this emphasis overshadows the genius of Landau's work in nuclear physics, quantum mechanics, superfluids (for which he won the 1962 Nobel Prize for Physics), superconductors, and astrophysics and his great influence on theoretical physicists. Photos. 207 pp.

Drell, Sidney D., and Sergei P. Kapitsa, eds. *Sakharov Remembered.* New York: American Institute of Physics, 1991.
Biography. Andrei Sakharov (1921-1989) was a man of sharp contrasts during his career: the father of the Russian hydrogen bomb and winner of the 1977 Noble Prize for Peace. He was exiled within the Soviet Union for espousing nuclear disarmament and often in trouble with the authorities for helping would-be Jewish emigrants. In this volume he is remembered

as an intellectual and moral leader, not only in the Soviet Union but in world science. The 29 essays are divided among three categories: personal reminiscences, including a biographical sketch by E. L. Feinberg and tributes by Kip S. Thorne and John Archibald Wheeler; scientific perspectives, with contributions by Andrei Linde and Hans Bethe and a reprinted article by Sakharov; and freedom of thought, which contains remarks collected from a symposium. Photos and drawings; end notes for each essay; index. 303 pp.

Dyson, Freeman. *Disturbing the Universe.* New York: Harper and Row, 1979.
Autobiography. Among the most renowned mathematical physicists of the twentieth century, Freeman Dyson wrote this book "to describe to people who are not scientists the way the main situation looks to somebody who is a scientist." The result, based upon 50 years of recollections, is an engrossing story. He describes the work he did with some of the greats of modern physics, such as Richard Feynman and Hans Bethe. The book opens with his childhood in England and his early mathematical talent, and then Dyson reminisces about coming to America, studying as a graduate student at Cornell University, and working on such diverse projects as arms control and the search for extraterrestrial intelligence while a professor of physics at the Institute for Advanced Study in Princeton. Richly cultured and literary, this book is for general readers, who will enjoy Dyson's lucid, amicable style. Bibliography; index. 283 pp.

'Espinasse, Margaret. *Robert Hooke.* Berkeley: University of California Press, 1956.
Scientific biography. Although understandable to general readers, this book concentrates on the ideas and research of Robert Hooke (1635-1703), and they were considerable. A genius of many interests and relentless self-promotion, Hooke was a star of the early Royal Society. 'Espinasse devotes chapters to the bitter rivalry between Hooke and Isaac Newton, Newtonian science in general, Hooke's *Micrographia* (optics, theory of combustion, petrifaction, and cellular structure of plants), horology (especially in relation to finding the

longitude), and his work as a surveyor and architect. She also discusses his social life and personal life. Photos and drawings; bibliography; end notes; index. 192 pp.

Feather, Norman. *Lord Rutherford.* Glasgow, Scotland: Blackie and Son, 1940; London: Priory Press, 1973.
Biography. Feather was a colleague of Ernest Rutherford (1871-1937) during the last decade of his life, and so Feather presents an insider's view of the father of nuclear physics. Written shortly after his death, the book views him as the patriarch of British science, as he was, and carefully presents the reasons for that stature. Feather begins with a chapter describing the scientific outlook when Rutherford began his career in order to show how much he helped change that outlook. Then Feather recounts Rutherford's youth in New Zealand, education at Cambridge University, professorships in Montreal and Manchester, and return to Cambridge as professor at the Cavendish Laboratory. Feather explicates Rutherford's experiments and discoveries for educated general readers. Photos; footnotes; index. 195 pp.

Fermi, Laura. *Atoms in the Family.* Chicago, Ill.: University of Chicago Press, 1954.
Biography. The author's memoirs of her life with her husband, Enrico Fermi. It is a strange, likable book, chatty about Fermi's colleagues and friends and clear about the basic elements of Fermi's work on the transmutation of elements and the first atomic reactor. Laura Fermi makes it clear, in spite of her adoration of her husband, that geniuses are not easy to live with. The story is often amusing and sometimes chilling, as when the Fermis had to flee fascist Italy to escape anti-Semitic persecution. This personal drama gives the book its richness, which is particularly true of the author's account of the Manhattan Project. Except for a short chapter on Fermi's childhood and school days, the book concentrates on the period between 1924 and 1953, shortly before he died. Photos. 267 pp.

Feynman, Richard P., as told to Ralph Leighton. *"Surely You're Joking Mr. Feynman!": Adventures of a Curious Character.* New

York: W. W. Norton, 1985; *"What Do You Care What Other People Think?": Further Adventures of a Curious Character*, 1988.
Autobiography. Taken together, these two books provide glimpses throughout Richard Feynman's life, and they are delightful, funny, lively, provocative reading. Leighton gives the readers the best of Feynman talking, and he was a famous talker. The subjects include much physics, explained non-technically, his childhood, his education at MIT and Princeton University, the Manhattan Project, his professorships at Cornell University and the California Institute of Technology, and his work on the committee investigating the space shuttle *Challenger* crash. The tone is usually irreverent; for instance, the section about his 1965 Nobel Prize for Physics and its effect on his life is titled "Alfred Nobel's Other Mistakes." Younger readers will find no better introduction to how it feels to be a scientist than these books, but all readers will appreciate their honesty and lack of pretense. Index in each book. 350 pp. and 255 pp.

Fölsing, Albrecht. *Albert Einstein: A Biography*. New York: Viking, 1997.
Biography. Fölsing is a science journalist in Germany, and his biography of Einstein understandably has a German slant and dwells on Einstein's life in Europe until 1932. His last two decades in America receive slightly less close treatment. Nonetheless, Fölsing's book, well translated by Ewald Osers, is a marvelous addition to the extensive literature about Einstein. It discusses his scientific achievements with unusual thoroughness and clarity for general readers, placing relativity and quantum theory in both their scientific and philosophical contexts. Fölsing also sheds light on the complexity of Einstein's character in his relationships with other physicists, his pacifism, and his support for Israel and Jewish refugees. Based on some newly released documents, the author also briefly discusses Einstein's lost daughter and his love affairs. Photos; bibliography; end notes; index. 882 pp.

Forsee, Aylesa. *Albert Einstein: Theoretical Physicist*. New York: Macmillan, 1963.
Biography. Suitable for high school students, this book tries to

prove that a reader does not have to be a physicist or mathematician to understand the basic ideas behind Albert Einstein's theories. Accordingly, Forsee dwells on the development of the special and general theories of relativity after a quick review of Einstein's family background and education. His contributions to quantum physics, political ideas, flight from Nazi persecution, and role in the making of the atomic bomb also receive attention. Photos and drawings; bibliography; glossary; end notes; index. 202 pp.

Frank, Philipp. *Einstein: His Life and Times*. New York: Knopf, 1953; Da Capo, 1989.
Biography. Frank was Albert Einstein's replacement as professor of theoretical physics in Prague and the two were friends. Thus, he has a wealth of anecdotes to tell about Einstein and close knowledge of his friend's effect on contemporary physicists. Although Frank says he writes for all those who want to understand the complex twentieth century, which to him Einstein best represents, the book is sophisticated in its presentation of physics concepts and the philosophy of science. Photos; index. 310 pp.

Freeman, Joan. *A Passion for Physics: The Story of a Woman Physicist*. Bristol, England: Institute of Physics Publishing, 1991.
Autobiography. A native Australian, Joan Freeman (b. 1918) went to England and trained as a physicist. She worked on the development of radar during World War II, completed her doctorate at Cambridge shorter afterward, and worked as an experimental physicist at the Cavendish Laboratory and the nuclear reactor facility at Harwell. Her book reveals an exciting era of physics, which she took part in only after overcoming resistance among some colleagues to having women in the field. Freeman concludes the book with an essay on women in physics. Throughout, she dwells on the fun of physics, which she believes more school girls should be encouraged to enjoy. Photos; index. 229 pp.

Frisch, Otto. *What Little I Remember*. Cambridge, England: Cambridge University Press, 1979.

Autobiography. Otto Frisch (b. 1904) grew up in Vienna—his aunt was Lise Meitner—and studied under or knew the pioneers of modern physics and mathematics, including Ernest Rutherford, Niels Bohr, Werner Heisenberg, Edward Teller, John von Neumann, and J. Robert Oppenheimer. Frisch coined the term *nuclear fission*, which he helped discover and develop during the Manhattan Project, moving to England afterward to continue research in nuclear physics. Not a skillfully written book, these memoirs still contain much of interest to students of twentieth century physics, especially in Frisch's penetrating comments about his contemporaries. Photos and drawings; bibliography; index. 227 pp.

Gamow, George. *My World Line.* New York: Viking, 1970.
Autobiography. Unfortunately, George Gamow (1904-1968) died before he could write a full-fledged autobiography. In this slim volume of memoirs, Gamow recalls his youth and education in Russia and tells a few stories about his scientific career in America. Gamow's literary flair and humor, evident in his many popular science works, come out strongly. The stories he tells are often hilarious and chilling at the same time. For instance, his father was Leon Trotsky's high school composition teacher, and they did not like each other very much. Although Gamow recalls his teachers and famous colleagues, such as Lev Landau and Niels Bohr, he seldom explains his own scientific work. For general readers. Photos; bibliography; index. 178 pp.

Garbedian, H. Gordon. *Albert Einstein, Maker of Universes.* New York: Funk and Wagnalls, 1939.
Biography. The combination of a florid style, a tone of adoration, and dramatic descriptive passages whose details are in part fictional makes this biography seem overwrought and sometimes even silly. Published before some of Einstein's most controversial scientific and political actions, the book is necessarily incomplete. It does, however, testify to the wide public curiosity about Einstein. Photos and drawings. 328 pp.

Gleick, James. *Genius: The Life and Science of Richard Feynman.* New York: Pantheon Books, 1992.

Biography. A brilliantly written account of one of America's most innovative and startling scientists, Richard Feynman (1918-1988), whom physicist Freeman Dyson once called "half genius and half buffoon." Among Gleick's many virtues as a biographer is his ability to place Feynman in his intellectual milieu and at the same time show the emotional forces influencing his work, such as his reverence for his father, his love for his critically ill first wife, and his worry over atomic weapons, which as a member of the Manhattan Project he helped invent. Another great virtue of this book is Gleick's clarity in explaining the abstruse physics that Feynman worked on, especially quantum electrodynamics (QED), for which he won a Nobel Prize in 1965. Gleick also recounts in depth Feynman's family life and his work investigating the space shuttle *Challenger* disaster. Photos; end notes; generous bibliography; index. 532 pp.

Goldman, Martin. *The Demon in the Aether: The Story of James Clerk Maxwell*. Edinburgh, Scotland: Paul Harris Publishing, 1983.
Biography. Goldman places James Clerk Maxwell (1831-1879) with Isaac Newton, Archimedes, and Albert Einstein as transcendent figures in the history of science. Certainly, as Goldman shows in detail, Maxwell's unification of electricity and magnetism in a single theory was a titanic achievement and set the stage for nearly all subsequent physics. The book is somewhat desultory in structure, concentrating on both Maxwell's life and career, especially as professor at the Cavendish Laboratory of Cambridge University, and his influence on scientific and intellectual history—a stimulating treatment for well-educated readers, although not necessarily scientists. Photos and drawings; end notes; index. 224 pp.

Goodchild, Peter. *J. Robert Oppenheimer, Shatterer of Worlds*. Boston, Mass.: Houghton Mifflin, 1981.
Biography. Goodchild passes over J. Robert Oppenheimer's considerable contributions to physics and astrophysics (including an early description of a black hole) in order to examine the enigma of his public persona—the father of the atomic bomb who was later denied a security clearance and in

effect tried for treason. Complemented by dozens of photographs, the book hastily covers Oppenheimer's youth, education, and early work at the University of California, Berkeley, and delves into the Manhattan Project in detail. Goodchild describes the work to invent the bomb, the people involved, and the politics, especially the antagonism that Oppenheimer (1904-1964) caused to some high-level members. Then he lays out the circumstances that led to the security hearing—the discovery that Oppenheimer's brother had been a member of the Communist Party and he himself had flirted with it, and his disagreement with military figures over the strategy of nuclear arms. The text, engagingly written for general readers, is troubling, for it presents an ugly picture of American politics and science. Photos and drawings; bibliography; index. 301 pp.

Gooding, David, and Frank A. J. L. James, eds. *Faraday Rediscovered*. New York: American Institute of Physics, 1989.
Scientific biography. The editors present 12 essays about the life and ideas of Michael Faraday from perspectives not taken in earlier biographies. The first four essays discuss Faraday's religion or tutelage at the Royal Institution, and the rest address his work in electricity, experimental style, field theory, and way of thinking. Photos; bibliography; end notes for each essay; index. 258 pp.

Gribbin, John, and Mary Gribbin. *Richard Feynman: A Life in Science*. New York: Dutton, 1997.
Biography. The authors believe that Richard Feynman was the best-loved scientist of modern times, yet that charisma has not been clear in previous biographies. Their book is intended to correct this deficiency. Furthermore, they insist that readers cannot understand Feynman's physics without understanding the kind of person he was. To Feynman physics was sheer joy. After setting the stage with a chapter explaining the state of physics before his career, the Gribbins convey this quality in recounting his boyhood and education, work in the Manhattan Project, contributions to quantum electrodynamics (QED) and superfluidity, Nobel Prize, role as a teacher at the California Institute of Technology, transactional interpretation

of quantum mechanics, and service on the panel investigating the explosion of the space shuttle *Challenger*. The Gribbins write movingly of Feynman, and the book explores his ideas with grace and clarity. Bibliography; end notes for each chapter; index. 301 pp.

Hartcup, Guy, and T. E. Allibone. *Cockcroft and the Atom*. Bristol, England: Adam Hilger, 1984.
Biography. In 1932, John Cockcroft (1897-1967) and Ernest Walton discovered that the atomic nucleus could be divided, a crucial step toward development of the atomic bomb more than a decade later. Cockcroft and Walton shared the 1951 Nobel Prize for Physics for their fission experiments. The authors follow Cockcroft's development from boyhood through his education and work at the Cavendish Physical Laboratory to reveal how he came to the discovery. Then they show Cockcroft during World War II and after, when as a scientist-diplomat of international reputation he worked for peace and improvement of underdeveloped countries. Photos; end notes; index. 320 pp.

Heilbron, J. L. *The Dilemma of an Upright Man: Max Planck as Spokesman for German Science*. Berkeley: University of California Press, 1986.
Biography. To Heilbron, the life of Max Planck (1858-1947) has elements of heroic tragedy. He began as a staunch defender of classical mechanics, specializing in entropy. Reluctantly, he proposed the quantum hypothesis and calculated the quantum of action, now symbolized by Planck's constant, h, which led to quantum mechanics and the end of the classical deterministic view of physics. He was a close friend of Albert Einstein and other German Jewish scientists but had to watch them flee the Nazis, to the impoverishment of German science. He was deeply patriotic but deplored the Nazis, who eventually drove him from his leadership role in the scientific community. Through Planck's many scientific triumphs and conflicts, Heilbron shows, he remained a man of model probity and humanity. Heilbron summarizes Planck's contributions to science and philosophy at length, although the bulk of the narrative concerns the politics of German science during

Planck's lifetime. An absorbing book for intellectually sophisticated readers. Photographs; footnotes; bibliography; index. 238 pp.

Heilbron, J. L. *H. G. J. Moseley: The Life and Letters of an English Physicist, 1887-1915*. Berkeley: University of California Press, 1974.
Biography. Henry Gwyn Jeffreys Moseley (1887-1915) died in the battle for Gallipoli during World War I, cutting short the career of the "most promising of all the English physicists of his generation." Heilbron shows that during a research career that lasted only 40 months Moseley accomplished an unusual amount in diverse fields, including X-ray spectography, radioactivity, periodic classification of the elements, and the theory of atomic structure. In fact, he helped develop the atomic models advanced by Niels Bohr and Ernest Rutherford. Heilbron's biography constitutes about half of this book, following Moseley from his birth in Weymouth in southern England through Eton, Oxford University, his research, military service, and death in battle. Heilbron's explanations of the scientific research require a basic understanding of physics to appreciate fully. The second half of the book contains Moseley's letters, both personal and scientific. Photos and diagrams; bibliography; footnotes; index. 312 pp.

Herivel, John. *Joseph Fourier: The Man and the Physicist*. Oxford, England: Clarendon Press, 1975.
Scientific biography. Joseph Fourier (1768-1830) was a leader of the French Revolution in his hometown, a high-ranking administrator for Napoleon Bonaparte, and a power in the French Academy. Herivel ably tells his remarkable story in the first six chapters of the book. The last four chapters take up Fourier's scientific work, for he was a leading theoretical physicist of his times: the equation of the motion of heat; heat flux in solid bodies; movement of heat through fluids; and analytical theory of heat, his most celebrated contribution. A knowledge of higher mathematics will help readers but is not necessary for most of the book. Photos; bibliography; end notes; index. 350 pp.

Hermann, Armin. *Werner Heisenberg, 1901-1976.* Bonn, Germany: Inter Nationes, 1976.
Biography. Hermann records the major steps in Werner Heisenberg's career and, to a lesser extent, his private life. Although Heisenberg was one of the architects of quantum mechanics, Hermann provides little explanation either of its physical principles in general or of Heisenberg's contributions. Rather, the focus is on academic politics and social relations among physicists as they produced a revolution in their discipline. The text is disjointed and underwritten, relying heavily on quotations from Heisenberg's letters and those of others. Photos; bibliography; end notes; index. 145 pp.

Highfield, Roger, and Paul Carter. *The Private Lives of Albert Einstein.* New York: St. Martin's Press, 1993.
Biography. Recently available letters and documents allow a fundamental reappraisal of Albert Einstein's character, the authors argue. They believe that the long-standing image of Einstein as the stoical scientific saint was a mask for a man of great passion and equally great failings as a father and husband. Based upon interviews with acquaintances and family members of Einstein, the book pays close attention to his first wife, Mileva Maric, and their two sons; his second wife, Elsa Einstein; and other women in his life. Such human relations, rather than science, are the subject of the book, which is deftly written for all readers, but especially those already familiar with Einstein's life. Photos; bibliography; end notes; index. 353 pp.

Hoffmann, Banesh, with Helen Dukas. *Albert Einstein: Creator and Rebel.* New York: Viking, 1972.
Biography. A loving portrait of Einstein by a former assistant, Hoffmann, and his long-time secretary and executor of the Einstein Archives, Dukas. Their thesis is that Einstein was a "profoundly simple man" with passionate curiosity and an unrivaled intuition about the beauty and structure of nature. The text develops this theme in the selection of events it relates and in its substantial but non-technical explanations of Einstein's scientific thought. His political and social ideas receive less thorough attention. Stylishly written, the book is

entertaining reading and presents the essence of Einstein's life work clearly. Many photos, drawings, and diagrams; index. 272 pp.

Infeld, Leopold. *Quest.* New York: Chelsea Publishing, 1980.
Autobiography. Infeld begins this dramatic, sad book with his reactions to news in 1939 that the capital of his homeland, Poland, had fallen to Nazi troops. He had become an American, having fled antisemitism in Europe and trying to start a new career. Then he backtracks and describes the ghetto in Krakow where he grew up and his efforts to escape it through education. Finally, he tells about his work in physics and his friendship with such figures as Albert Einstein, with whom he collaborated on a classic popular physics book. A pensively written, moving book for all readers. 361 pp.

Ipsen, David C. *Archimedes: Greatest Scientist of the Ancient World.* Hillside, N.J.: Enslow, 1988.
Biography for readers ten years and older. Ipsen describes the life and ideas of Archimedes (287-212 B.C.E.) simply and clearly. He presents the physical basis of buoyancy and the importance of squaring the circle especially well for young readers. The book also affords some understanding of the role of scientific knowledge in the ancient world. Photos, drawings, and diagrams; glossary; short bibliography; index. 64 pp.

Jaffe, Bernard. *Michelson and the Speed of Light.* New York: Doubleday, 1960; Westport, Conn.: Greenwood Press, 1979.
Biography-popular science. One of the most famous and consequential of American experiments, the unsuccessful attempt by Albert A. Michelson (1852-1931) and Edward W. Morley to detect the ether showed that something was basically wrong with the nineteenth-century view of the universe. The experiment set the stage for Albert Einstein's formulation of relativity. Concentrating on Michelson, the first American to win a Nobel Prize, Jaffe tells the story of a triumphant scientific career. He also explains the reasons for the ether theory and Michelson's measurements of the speed of light, as well as his interferometer experiment with Morley. That they failed to

prove the ether's existence disappointed Michelson, and he did not suspect its revolutionary significance. In fact, he long resisted the special theory of relativity. Jaffe's account is a penetrating view of classical physics and how it prepared for new ideas at the turn of the twentieth century. A smoothly readable account for non-scientists. Photos and diagrams; bibliography; index. 197 pp.

Johnson, V. A. *Karl Lark-Horovitz*. Oxford, England: Pergamon Press, 1969.
Scientific biography. Johnson devotes a short section to the background, education, and teaching career of Karl Lark-Horowitz (1892-1958) and his World War II work developing radar, about 50 pages. The rest of the book concerns his research in X-ray and electron diffraction and solid state physics, especially the behavior of such semiconductors as germanium. The scientific section requires an understanding of physics from readers. Photos and diagrams; end notes for each chapter; index. 289 pp.

Kargon, Robert H. *The Rise of Robert Millikan*. Ithaca, N.Y.: Cornell University Press, 1982.
Biography. Intending this book for general readers, Kargon says it is not a full biography; rather, he uses the particularly American career of Robert Millikan (1868-1953) to consider how science took such a strong hold on American culture during the years when Millikan flourished and almost on his own represented science in this country. His early experiments helped confirm theories about the structure of the atom and behavior of electrons, which brought him the 1923 Nobel Prize for Physics, only the second for an American. He led the California Institute of Technology into the front ranks of scientific centers, and late in his career he pursued an incorrect theory of cosmic rays that tarnished his reputation. After a chapter about Millikan's youth and education, Kargon analyzes Millikan's career for its broader significance to American culture. Photos; bibliographic essay; end notes; index. 203 pp.

Kedrov, F. B. *Kapitsa: Life and Discoveries*. Moscow: Mir Publishers, 1984.

Biography. Pyotr Leonidovich Kapitsa (1894-1984) won a share of the 1978 Nobel Prize for Physics for his research in superfluidity and was famous as a student of Ernest Rutherford at the Cavendish Laboratory and a very young member of the Royal Society. When he returned to Russia from his studies in England, the authorities never again permitted him to leave. After setting up his own Institute for Physical Problems in Moscow, Kapitsa often defied the authorities, refusing to help Stalin build the hydrogen bomb, for which he was placed under house arrest, and helping imprisoned colleagues, such as Lev Landau. Kedrov emphasizes Kapitsa's scientific and administrative work in this brief book, relating his youth, education at Cambridge, return to Moscow, efforts during World War II, and teaching. Photos; bibliography; end notes. 199 pp.

Kilmister, Clive William, ed. *Schrödinger: Centenary Celebration of a Polymath*. Cambridge, England: Cambridge University Press, 1987.
Scientific biography. This collection of essays from a conference honoring Erwin Schrödinger (1887-1961) is almost strictly for physicists. Most of the essays analyze his contributions to physics, such as wave mechanics, and involve sophisticated concepts and higher mathematics. General readers, however, can understand the essays concerning his years in Ireland and his philosophy. Among the essays' authors are John S. Bell, Abdus Salam, Stephen W. Hawking, and Linus Pauling. End notes for each essay; index. 253 pp.

Kipnis, A. Y., B. E. Yavelov, and J. S. Rowlinson. *Van der Waals and Molecular Science*. Oxford, England: Clarendon Press, 1996.
Scientific biography. Johannes Diderik van der Waals (1837-1923) championed the theory that atoms, molecules, and their interactions lie behind chemistry and much of physics, a view that prevailed after a long struggle. He thereby fathered modern physical chemistry, and his success brought him the 1910 Nobel Prize for Physics. This translation of the 1985 Russian edition reappraises Van der Waal's contribution, which the authors believe has been underappreciated. They do so by examining Van der Waal's family background and youth in

Leiden, education, thesis on the behavior of gases, professorship at the University of Amsterdam, research in molecular physics and physical chemistry, membership in the Royal Academy of Sciences, and relation to Russian science. The discussion is often technical, requiring the reader to understand the basics of chemistry and physics. The book provides much information about a neglected major figure. Photos and diagrams; bibliography; end notes; indexes. 313 pp.

Klein, Martin J. *Paul Ehrenfest*. Amsterdam, The Netherlands: North-Holland Publishing, 1970.
Scientific biography. Paul Ehrenfest exerted great influence on physics during the early twentieth century as a professor at the University of Leyden. His work on quantum mechanics and the adiabatic principle places him in the first rank of theorists, and the outgoing Ehrenfest was friends with many of them. Albert Einstein and he were particularly close, and this book concludes with an exposition of that friendship. Klein also discusses Ehrenfest's childhood, education, dissertation on mechanics, theoretical ideas, and teaching. Klein makes use of advanced mathematics in his explanations and assumes that his readers understand the principles of physics. Photos; footnotes; index. 330 pp.

Kragh, Helge. *Dirac*. Cambridge, England: Cambridge University Press, 1990.
Scientific biography. Kragh presents the essential personal facts in the somewhat cloistered life of Paul Adrien Maurice Dirac (1902-1984), but the bulk of the text explores his physical theories. They are conceptually intricate and pervasively influential in modern physics, especially his work on electron dynamics and quanta. Dirac shared the 1933 Nobel Prize for Physics with Erwin Schrödinger for his mathematical treatment of quantum mechanics. One also finds discussion of such topics as cosmology, the philosophy of physics, and mathematical beauty. Readers must understand advanced physics and mathematics to profit from Kragh's presentation. Photos, drawings, and diagrams; bibliography; end notes; indexes. 389 pp.

Kursunoglu, Behram N., and Eugene Wigner, eds. *Reminiscences About a Great Physicist: Paul Adrien Maurice Dirac*. Cambridge, England: Cambridge University Press, 1987.
Scientific biography. This book contains 22 essays about the life and ideas of Paul Dirac by those best in a position to understand: his peers in physics, among whom, in addition to Wigner, are five Nobel laureates. The first of the book's three sections starts with an essay by Dirac's wife; the rest of the authors recall stories from his career. The second section's essays comment upon his physics, including magnetic monopoles and quantum field theory. The final section discusses Dirac's influence on other physicists and other disciplines. Many essays in the last two sections require an advanced knowledge of physics to comprehend. Photos and diagrams; end notes for each essay; indexes. 297 pp.

Kurylo, Friedrich, and Charles Susskind. *Ferdinand Braun, a Life of the Nobel Prizewinner and Inventor of the Cathode-Ray Tube*. Cambridge, Mass.: MIT Press, 1981.
Biography. A translation of the 1965 German edition. For inventing the cathode-ray tube, precursor of television, the German physicist Ferdinand Braun (1850-1918) shared the 1908 Nobel Prize for Physics with Guglielmo Marconi. As the authors show, however, he did far more in physics, such as discovering the rectifier effect, contributing to thermodynamics, and fabricating magnetic compounds. They discuss his childhood, university years, teaching positions at Strasbourg University and other institutions in Germany, and participation in developing electrical companies. Photos and diagrams; end notes; indexes. 289 pp.

Latil, Pierre de. *Enrico Fermi: The Man and His Theories*. New York: Paul S. Eriksson, 1966.
Biography. A translation from the French edition, this book exalts Fermi relentlessly. The narrative is fragmented and heavily indebted to Laura Fermi's biography of her husband (see above). The book's greatest strength lies in its clear, detailed explanations of Fermi's many scientific contributions. Appendices contain stories about Fermi by other writers and some extracts of Fermi's writings, including his speech

accepting the Nobel Prize for Physics in 1938. Photos; bibliography; index. 178 pp.

Lindsay, Robert Bruce. *Julius Robert Mayer, Prophet of Energy.* Oxford, England: Pergamon Press, 1973.
Scientific biography. In the first of this book's three parts Lindsay provides a biographical sketch of Julius Robert von Mayer (1814-1878), one of the founders of thermodynamics and biophysics. The second part reviews Mayer's research into the nature of energy and heat. The third part reprints five of Mayer's articles with editorial commentary. Lindsay writes primarily for scientists and historians. Bibliographies for the first two sections; index. 238 pp.

Lindsay, Robert Bruce. *Lord Rayleigh—The Man and His Work.* Oxford, England: Pergamon Press, 1970.
Scientific biography. A short biographical sketch of John William Strutt, Lord Rayleigh (1842-1919), prefaces a brief review of his contributions to physics, a wide range of work that included acoustics, optics, electrical standards, blackbody radiation, and the discovery of argon. The discovery brought him the 1904 Nobel Prize in Physics. The largest part of this book comprises reprints of Rayleigh's major articles, most of which require a professional competency in physics to grasp. Bibliography; index. 251 pp.

Livingston, Dorothy Michelson. *The Master of Light.* New York: Charles Scribner's Sons, 1973.
Biography. A loving portrait of Albert A. Michelson by his daughter. Michelson, with Edward Morley, performed an experiment famous for its failure to detect the motion of ether. The experiment led indirectly to the special theory of relativity, which Michelson long hesitated in accepting. For the work, Michelson received the 1907 Nobel Prize for Physics, the first for an American. Livingston tells of her father's lifetime obsession with light—measuring its speed and learning its principles—and explains the physics for general readers who (as she was when she began writing the book) are new to physics. But most of the book concerns Michelson's personal life and career, from his birth in Poland to his professorship at

the University of Chicago and friendships with the leading American physicists. Photos, drawings, and diagrams; bibliography; end notes; index. 376 pp.

Maddison, R. E. W. *The Life of the Honourable Robert Boyle*. London: Taylor and Francis, 1969.
Biography. Maddison takes the life of Robert Boyle (1627-1691) period by period: childhood, Eton, the Grand Tour (1627-1644); the Stalbridge period (1645-1655); the Oxford period (1656-1668); and the London period (1668-1691). He explains the physical ideas of this great Anglo-Irish scientist, including the air pump and work on gases, and his participation in the fledgling Royal Society. Maddison quotes extensively from Boyle's own works and reprints his will in an appendix. Photos and drawings; index. 332 pp.

Mehra, Jagdish. *The Beat of a Different Drummer: The Life and Science of Richard Feynman*. Oxford, England: Oxford University Press, 1994.
Scientific biography. Mehra presents the substance of the scientific achievements of Richard Feynman at length throughout the book. These explanatory sections require advanced training in physics and mathematics to appreciate fully. Among the achievements are Feynman's considerable contributions to quantum mechanics, quantum electrodynamics (QED, for which he shared a Nobel Prize in 1965), superfluid theory, the theory of the weak interaction, gravitation, and fundamental particle research. Mehra shows well why Feynman was known as the physicist's physicist. He also recounts Feynman's private life and relations with colleagues. He based large sections of these passages on extensive interviews with Feynman. Unlike the sections about physics, the information about Feynman's life is neither adroitly presented nor particularly penetrating. Still, there are many remarkable anecdotes about Feynman's family, fellow physicists, and his role in investigating the space shuttle *Challenger* disaster. Photos; chapter end notes; index. 630 pp.

Michelmore, Peter. *Einstein, Profile of the Man*. New York: Dodd, Mead, 1962.

Biography. Michelmore interviewed Hans Albert Einstein, elder son of Albert Einstein, extensively about his father, the first biographer to do so. The information gathered thereby, Michelmore intimates, allows him to probe for the man behind the popular myth of Einstein as the great sage in an ivory tower spinning abstruse theories about the universe. Michelmore opens the book with general explanations of relativity and the circumstances in which Einstein formulated the general theory. Then Michelmore backtracks to Einstein's childhood and sketches his life chronologically. Despite the interview, much of the story comes from published sources. Index. 269 pp.

Michelmore, Peter. *The Swift Years*. New York: Dodd, Mead, 1969.
Biography. Michelmore's subject is J. Robert Oppenheimer, scientific director of the Manhattan Project, leader of the first theoretical physics center in the United States, and pioneer in black hole physics. Oppenheimer was a complex man; unfortunately, this book does not provide much insight into his character. Anecdotal in style, it dwells on how eclectic Oppenheimer's intellect was, defends his political allegiance against those who thought he was a Soviet agent, shows he loved dramatic poses, and leaves the reader wondering how anyone could like him. Very little of Oppenheimer's work in physics receives close attention. Photos; bibliography; index. 273 pp.

Moore, Ruth. *Niels Bohr: The Man, His Science, and the World They Changed*. New York: Alfred A. Knopf, 1966.
Biography. Written for non-scientists, this book tries to capture the many facets to the ideas and personality of Niels Bohr, widely acknowledged as the father of quantum mechanics. Moore explains the general principles of Bohr's revolutionary work in atomic physics, quantum mechanics, and nuclear physics, but she goes into more depth about Bohr's political ideas for a peaceful world, philosophy based on complementarity, friendship with all the great physicists from the first half of the twentieth century, institute in Copenhagen, Denmark, friendly disagreements about quan-

tum mechanics with Albert Einstein, and participation in the Manhattan Project. She also shows Bohr's gentleness and humanity in his dealings with his family and friends. This is a readable, informative book, well suited as a first book for readers new to the bewildering world of modern physics, because Moore captures the excitement, confusion, and emotional impact of discoveries in the atomic realm. Photos and diagrams; bibliographical note; index. 443 pp.

Moore, Walter. *Schrödinger, Life and Thought.* Cambridge, England: Cambridge University Press, 1989.
Scientific biography. Erwin Schrödinger is best known for formulating the equations for wave mechanics, which describes the wave properties of both matter and energy. His contribution to the development of quantum mechanics is central, even though he disagreed with Niels Bohr and Werner Heisenberg about how to interpret the physical bases of quanta. Moore presents the philosophical, scientific, and mathematical aspects of Schrödinger's achievements and controversies clearly and in depth. In passing he supplies valuable mini-biographies of Schrödinger's scientific forebears, such as Ludwig Boltzmann. Moore is much less clear about Schrödinger's personal life, which involved many love affairs. Moore often applies psychological theories ad hoc to shed light on his subject's psyche, a procedure which confuses. Readers must understand calculus to appreciate the text fully. Photos and diagrams; end notes; indexes. 513 pp.

More, Louis Trenchard. *The Life and Works of the Honourable Robert Boyle.* New York: Oxford University Press, 1944.
Biography. Writing for historians primarily, More argues that Robert Boyle was a signal transition figure in an age of transition from the medieval period to the early modern period and the origins of science. Accordingly, More discusses Boyle's work as an alchemist as well as his analysis of the behavior of gases. More also examines Boyle's theological beliefs and the New Philosophy he absorbed at Oxford. Throughout the narrative, More places Boyle's character, ideas, and career in the context of the beliefs and institutions of seventeenth-century Great Britain. Portrait of Boyle; end notes; index. 313 pp.

Pais, Abraham. *Einstein Lived Here.* New York: Oxford University Press, 1994.
Biographical essays. A companion volume to *"Subtle Is the Lord..."* (see second entry below). The essays provide information based on new sources or matters that did not fit into the full biography: Einstein's relationships with his first wife and their daughter; Einstein's friendship with Niels Bohr; Louis De Broglie and Einstein; how he got the Nobel Prize; and his discussions with the Indian poet Rabindranath Tagore. Other sections of the book contain excerpts from news stories and letters or explanations of Einstein's philosophical ideas and relativity. Enjoyable reading, containing far less technical detail about physics than Pais' *"Subtle Is the Lord..."*. Photos; end notes for each essay; index. 282 pp.

Pais, Abraham. *Niels Bohr's Times, in Physics, Philosophy, and Polity.* Oxford, England: Clarendon Press, 1991.
Scientific biography-history. This blend of science explication, biography of Bohr, and Pais' reminiscences is exhaustive and demanding reading for non-scientists because it contains a moderately high level of technical detail and mathematics. Pais knew Bohr late in his life and became friends with his son, so some of his information is fresh and firsthand. After explanatory background about relativity and the roots of quantum physics, Pais discusses the sweeping range of topics to which Bohr contributed directly or influenced others to contribute at his prestigious institute in Copenhagen, Denmark: the model of the atom, quantum mechanics and the "Copenhagen interpretation" of it fashioned in large part thanks to Bohr, quantum electrodynamics, particles, fields, the neutrino, and fission. Pais also recounts Bohr's relationship with his longtime friend and intellectual opponent, Albert Einstein, and discusses the philosophical extensions of Bohr's ideas, particularly complementarity. Photos; indexes. 565 pp.

Pais, Abraham. *"Subtle Is the Lord..."; The Science and Life of Albert Einstein.* New York: Oxford University Press, 1982.
Scientific biography. An eminent physicist who knew Albert Einstein at Princeton, Pais is an almost ideal explicator of the theories, and this book, a minor classic in its genre, is a lively,

absorbing account of the concepts of physics Einstein encountered as a young man, how he changed them, and his legacy to science. It is nevertheless a difficult book, fully comprehensible only to readers who understand basic physics and calculus. He mixes personal reminiscences, biography, science history, and sophisticated science explanations in each of seven sections that follow Einstein's life and career. The first six sections regard Einstein's youth and the background to relativity and quantum mechanics; statistical physics; special relativity; general relativity; his attempts to formulate a unified field theory; and his philosophical objections to quantum mechanics. The final section is a summation and general assessment of Einstein's overall contribution to physics. Photos and diagrams; end notes; indexes. 552 pp.

Peat, F. David. *Infinite Potential: The Life and Times of David Bohm.* Reading, Mass.: Addison-Wesley, 1997.
Biography. David Bohm (1917-1992) was a maverick physicist and philosopher, something of a tragic story. He helped develop the foundations of plasma and solid-state physics and wrote an influential book on quantum mechanics. When he reinterpreted quantum mechanics with his own "hidden variable" approach he intrigued Albert Einstein, Wolfgang Pauli, and Niels Bohr but others largely ignored or derided him, including his mentor, J. Robert Oppenheimer. Passionate about ideas, emotionally unstable, and politically unorthodox, Bohm ran afoul of Congress in the late 1940's for being a Marxist and Communist Party member. He fled to Brazil and ended up an expatriate in England. By then he had borrowed from the philosophies of Georg Hegel and Jiddu Krishnamurti to extend his physical ideas into a theory of "implicate order" that holds the material world as an expression of deeper relationships. While Peat effectively outlines Bohm's ideas for a general audience, implicate order remains elusive in this book, as does Bohm's personality. Photos; end notes; index. 353 pp.

Reef, Catherine. *Albert Einstein, Scientist of the 20th Century.* Minneapolis, Minn.: Dillon Press, 1991.
Biography for readers eight years and older. The book begins

with a summary of Albert Einstein's scientific achievements and beliefs. Then Reef starts the main narrative with a brief anecdote about Einstein's first visit to America in 1921. She backtracks to his boyhood and continues with his theory of relativity, professorships, troubles as a famous Jew in Nazi Germany, immigration to the United States, role in the atomic bomb project during World War II, and involvement with Zionism. Photos; glossary; index. 63 pp.

Rigden, John S. *Rabi, Citizen and Scientist*. New York: Basic Books, 1987.
Biography. Isador Isaac Rabi (1898-1988) won the Nobel Prize for Physics for his experimental and theoretical investigations of the magnetic properties of the atomic nucleus. Along with J. Robert Oppenheimer, he was among the generation of American physicists who trained in the great European scientific institutes and returned to America in the 1920's and 1930's determined to make physics in the United States equal to or better than that in Europe. More than any other single person, Rabi has been associated with that successful effort, and he became known as the dean of American physics. The author extols this development while tracing Rabi's career as a scientist, teacher, and government advisor concerning atomic power plants and weapons programs. Photos; end notes; index. 302 pp.

Rosental, Stefan, ed. *Niels Bohr: His Life and Work as Seen by His Friends and Colleagues*. Amsterdam, The Netherlands: North-Holland Publishing, 1967.
Biography. The Danish physicist Niels Bohr was a dominant figure in twentieth-century physics as the father of quantum mechanics and in the philosophy of science as the originator of the complementarity principle. He seems to have known everyone of importance in science and government, and they all held him in respect and affection. That at least is the view one gets from this collection of biographical essays, reminiscences, and philosophical interpretations of Bohr's life and ideas. Among the 20 contributors are Bohr's sons, Léon Rosenfeld, Werner Heisenberg, Paul A. M. Dirac, Abraham Pais, and former Danish prime minister Viggo Kampmann.

The authors write warmly, often movingly, of Bohr, covering his life, work at his theoretical physics institute, relations with other scientists, efforts for atomic energy and nuclear disarmament, and philosophy. Pleasant reading throughout, the book offers perspectives on Bohr rather than a coherent view of him but succeeds in its general purpose, which is to show that he was a man of great humanity and intelligence. Photos and drawings. 355 pp.

Rouzé, Michel. *Robert Oppenheimer, the Man and His Times*. New York: Paul S. Eriksson, 1965.
Biography. Rouzé believes J. Robert Oppenheimer symbolizes the difficult relation of the scientist to modern society and explores the theme in light of the potential modern science has, he claims, to destroy civilization. Accordingly, the author is less concerned about Oppenheimer as a person than as an opportunity for critiquing the culture of physics in the twentieth century. Rouzé addresses Oppenheimer's childhood and work in quantum physics in two short chapters, then devotes five chapters to the atomic bomb and its consequences to politics. The book concludes with transcripts of a lecture about knowledge by Oppenheimer, two interviews with him, an essay summarizing his work in physics, and a brief history of the Institute for Advanced Study at Princeton, where Oppenheimer served as director. Photos; glossary; bibliography; index. 192 pp.

Sagdeev, Roald Z. *The Making of a Soviet Scientist: My Adventures in Nuclear Fusion and Space from Stalin to Star Wars*. New York: John Wiley and Sons, 1994.
Autobiography. Roald Sagdeev (b. 1932) is an expert in plasma physics who changed to space science and became director of the Space Research Institute of the Soviet Union. Subsequently he moved to the United States to head the East-West Center for Space Science at the University of Maryland. He is a charming, witty writer who supplies vivid portraits of some of Russia's greatest physicists, his mentors and colleagues: Lev Landau, Peter Kapitsa, and Andrei Sakharov, for instance. He spends little time on the details of science; instead, he relates the politics of Soviet science in detail, from

the era of Stalin to that of Mikhail Gorbachev, whom Sagdeev occasionally advised on space projects and nuclear disarmament. His anecdote-filled narrative is sometimes chilling and often hilarious but finally poignant, since modern Russia has lost many of its brightest scientists, as Sagdeev ruefully notes. Glossary of names; index. 339 pp.

Sayen, Jamie. *Einstein in America.* New York: Crown Publishers, 1985.
Biography. Although Sayen allots two chapters to Albert Einstein's earlier years and European career, he devotes the book to the period between 1933 and 1955, when Einstein lived in the United States. The book provides a sophisticated look at Einstein's scientific and political involvements. Sayen contends that both sprang from Einstein's belief that everyone has an ethical imperative to seek the truth, that the truth resides in nature, and that one best pursues truth by abandoning self-absorption. Sayen recounts Einstein's public and private life in considerable detail, especially his involvement in internationalist movements, nuclear arms control, Zionism, and pacifism and his stand against McCarthyism. He draws on interviews with Einstein's stepdaughter, secretary, and other close associates. Photos and drawings; end notes; index. 340 pp.

Seeger, Raymond J. *Benjamin Franklin: New World Physicist.* Oxford, England: Pergamon Press, 1973.
Scientific biography. Seeger has performed a service: In the many biographies of Benjamin Franklin (1706-1790), it is difficult to find substantive discussions of his science, since most writers concentrate on his vast political importance. Seeger does the opposite. In the brief first section of the book he describes Franklin's involvement in printing, philosophy, and the American Revolution. The second section of the book, five times as long, concerns Franklin's study of evaporation, thermal absorption, behavior of fluids, electricity, and weather, as well as his invention of a new type of fireplace and musical glasses. Much of the discussion involves long passages excerpted from Franklin's letters. Drawings; bibliography; indexes. 190 pp.

Segrè, Claudio. *Atoms, Bombs and Eskimo Kisses.* New York: Viking, 1995.

Autobiography. What is it like to be the son of a legendary physicist, a Nobelist and member of the Manhattan Project? For those who have wondered, Claudio Segrè supplies one son's answer in this long, passionate, brooding book about his relation to his father, Emilio Segrè (1905-1989). The book contains little about the father's discoveries in physics—he is best known as co-discoverer of the antiproton—but there are many anecdotes about famous colleagues, such as Enrico Fermi, and a chapter about daily life in Los Alamos during World War II. Photos; end notes. 287 pp.

Segrè, Emilio. *Enrico Fermi, Physicist.* Chicago, Ill.: University of Chicago Press, 1970.

Scientific biography. A Nobel laureate in physics himself, Segré anatomizes the scientific achievements of his friend and mentor, Enrico Fermi (1901-1954). He also attempts to satisfy the curiosity of readers about the character of Fermi, although he refers them elsewhere for information about Fermi's private life. Segrè writes meticulously of Fermi's work on artificial radioactivity, in which Segrè participated, his building of the first atomic pile, and his contribution to the Manhattan Project. The book is a valuable resource, especially since Segrè explains scientific matters without employing technical jargon or mathematics beyond the grasp of the average reader. Appendices contain some of Fermi's correspondence and lectures. Photos; bibliography; end notes; index. 276 pp.

Sime, Ruth Lewin. *Lise Meitner: A Life in Physics.* Berkeley: University of California Press, 1996.

Biography. A well-presented account of a physicist whose career was scarred by tragedy. Like Marie Curie, Lise Meitner (1878-1968) overcame the gender barrier to earn a doctorate in physics and then, after long and brilliant labor, won a position at the prestigious Kaiser Wilhelm Institute for Chemistry. There she helped Otto Hahn discover nuclear fission. Shortly afterwards, the Nazi government forced her to flee to Sweden because she was Jewish. Although Hahn won a Nobel Prize for the discovery, Meitner was described as only a helper, and according to Sime, many physicists and historians regard the snub of her as one of the greatest injustices of modern science. Sime carefully describes Meitner's considerable achievements and her friendship with leading physicists of the twentieth century. Photos; bibliography; end notes. 526 pp.

Simon, Sheridan. *Stephen Hawking: Unlocking the Universe.* Minneapolis, Minn.: Dillon Press, 1991.
Biography for readers 14 years and older. The author limns a heroic portrait of Hawking as both unmatched in his intellectual power and unbowed by his struggle with Lou Gehrig's disease (amyotrophic lateral sclerosis). As well as outlining Hawking's life and academic career, Simon explains fundamental ideas in black hole physics and cosmology. Photos; glossary; bibliography; index. 112 pp.

Smith, Crosbie, and M. Norton Wise. *Energy and Empire.* Cambridge, England: Cambridge University Press, 1989.
Biography. The authors supply an exhaustive study of William Thomson, Lord Kelvin (1824-1907). They find that his brilliant advances in mathematical physics were inextricably related to his much-criticized work in industrial technology. Because information on Thomson's private life is scarce, the authors devote most of the study to his public and scientific careers. They divide the book into four thematic sections: his youth, family, and education; mathematical physics (including Kelvin's work in mechanics, electrodynamics, and thermodynamics); cosmological and geological theories; and practical activities in industry (for example, his nautical compass and the transatlantic cable). Some sections require the reader to know advanced mathematics. Photos and drawings; footnotes; bibliography; index. 866 pp.

Sugimoto, Kenji. *Albert Einstein: A Photographic Biography.* New York: Schocken Books, 1989.
Biography. On an official form, Albert Einstein once filled in the space for occupation with "photographer's model," perhaps as much from frustration as mischievousness. Certainly, he was the darling of news photographers, as this collection of Einstein memorabilia testifies. Sugimoto has assembled more

than 400 photographs of Einstein, as well as of friends, family, colleagues, teachers, enemies, documents, letters, scientific journal title pages, and significant political events. Sugimoto also recounts, briefly and adoringly, the main events and interests in Einstein's life. For Einstein devotees this book is a treat. Drawings and graphics; bibliography; index. 202 pp.

Sykes, Christopher. *No Ordinary Genius: The Illustrated Richard Feynman*. New York: W. W. Norton, 1994.
Biographical reminiscences. Sykes presents memories of friends, family members, and colleagues about Richard Feynman and excerpts from his own reminiscences. Among those quoted are Hans Bethe, Murray Gell-Mann, Freeman Dyson, Marvin Minsky, and John Archibald Wheeler. Drawn from a television documentary, the book reads like captions to Feynman's life, which included work for the Manhattan Project and the Nobel Prize for Physics. Many photos; end notes; short bibliography; index. 272 pp.

Thomson, Joseph John. *Recollections and Reflections*. London: G. Bell and Sons, 1936; New York: Arno Press, 1975.
Autobiography. J. J. Thomson (1856-1940), 1906 Nobel laureate for physics for discovering the electron, became an elder statesman of science in England as professor of experimental physics at Cambridge University's Cavendish Laboratory. In this delightful book, he recalls his boyhood, university days, tenure at the Cavendish, research, travels on lecture tours, and friendships with such scientists as Ernest Rutherford and Lord Rayleigh. He assumes that readers know the basic principles of physics and can follow his occasional use of mathematics. Photos and diagrams; indexes. 451 pp.

Townes, Charles L. *Making Waves*. Woodbury, N.Y.: American Institute of Physics, 1995.
Scientific memoirs. This volume collects 15 articles and lectures by Charles L. Townes, who shared the 1964 Nobel Prize for Physics for developing the maser and laser. The essays reflect upon physical principles, engineering problems, his role in quantum electronics, the maser and laser, microwave spectroscopy, radio astronomy, technology in society, and sci-

ence education. He also ruminates upon the nature of science, religion, and life as a physicist. Photos and diagrams; index. 210 pp.

Vallentin, Antonina. *The Drama of Albert Einstein*. Garden City, N.Y.: Doubleday, 1954.
Biography. The drama referred to in the title is the persecution that Albert Einstein faced in Germany because he was Jewish and his flight to escape the Nazis. Vallentin was a close friend of Einstein's second wife and so was well informed of the episode. The book also covers Einstein's youth, education, theories, and rise to fame, but Einstein's self-exile and his arrival in the United States take up most of the narrative. A skillful writer, Vallentin tells the story with animation. The book, like most of the early biographies, idolizes Einstein. Photographs; index. 312 pp.

Venkataraman, G. *Journey into Light: Life and Science of C. V. Raman*. Bangalore, India: Indian Academy of Sciences, 1988.
Scientific biography. Chandrasekhara Venkata Raman (1888-1970) won the 1930 Nobel Prize for Physics for research into how light is scattered when directed through a transparent substance. The "Raman effect" brought him notice worldwide, and he was knighted for his achievement and became a member of the Royal Society. In India he was a national hero, as the author relates in this substantial review of Raman's life and work at the research institute in Bangalore that he founded. A physicist himself, Venkataraman reviews Raman's theories in technical detail; only readers with an understanding of advanced physics will appreciate the book fully. Photos and diagrams; bibliography; end notes; indexes. 570 pp.

Weisskopf, Victor. *The Joy of Insight: Passions of a Physicist*. New York: Basic Books, 1991.
Autobiography. Weisskopf (b. 1908) was among the generation of physicists trained immediately after the invention of quantum mechanics in the late 1920's. Viennese by birth, he studied there and in Germany, Denmark, and England, where he met and worked with many of the leading physicists of the era, including Niels Bohr, Max Born, Werner Heisenberg, and

J. Robert Oppenheimer. He contributed to quantum mechanics, worked on the Manhattan Project, helped develop nuclear physics, and from 1961-1966 directed the European Center for Nuclear Research (CERN), then housing the most powerful particle accelerator in the world. This is a plainly written autobiography full of anecdotes of brilliant colleagues and insights into the aesthetic motives for pursuing physics. Weisskopf discusses physics principles at a level suitable to general readers with little science knowledge and displays a broad interest in the humanities as well. Photos; index. 336 pp.

Wheeler, Lynde Phelps. *Josiah Willard Gibbs*. New Haven, Conn.: Yale University Press, 1952.
Biography. Josiah Willard Gibbs (1839-1903) was the first American to achieve international stature as a theoretical physicist. His development of thermodynamics and statistical mechanics, which helped inaugurate physical chemistry, especially won the respect of such European scientists as James Clerk Maxwell, Lord Kelvin, Joseph John Thomson, and Ludwig Boltzmann, but he was largely unknown in the United States. Wheeler, a student of Gibbs at Yale, provides a somewhat restrained biography of his mentor, portraying him as a retiring gentleman scholar both abstracted from society in his pursuit of nature's secrets and cordial to his students and colleagues. Wheeler also explains Gibbs' great innovations in mathematical physics in sufficient detail to satisfy the determined college-educated reader. Overall, however, the book gives only a sketchy view of Gibbs' character and intellect. Photos; bibliography; index. 270 pp.

White, Michael, and John Gribbin. *Einstein: A Life in Science*. New York: Dutton, 1993.
Biography. White and Gribbin find that the increasing focus on Albert Einstein's character, especially the supposed bad behavior toward his family and his womanizing, have been absurdly exaggerated, an effect of a strident demythologizing of the scientist after his image had been too closely guarded by zealous literary executors. The authors find Einstein to have been a man of moral weaknesses, in truth, but it is his scientific genius, not his faults, that deserve study and admi-

ration. Accordingly, while they mention the controversies surrounding Einstein's personal life, they focus on relativity, quantum physics, and other aspects of Einstein's pioneering ideas, as well as his political beliefs. This is a clear, engaging introduction to Einstein for readers who have not yet read a biography of him. Bibliography; end notes; index. 279 pp.

White, Michael, and John Gribbin. *Stephen Hawking: A Life in Science.* New York: Dutton, 1992.
Biography. Pleasant reading, but a somewhat disconcerting book. The authors explain basic ideas in black hole physics and cosmology so that the contributions of Stephen W. Hawking (b. 1942) are broadly clear; these explanations are unexceptionable, if superficial. Most of the text, however, is devoted to Hawking's life, especially his meteoric rise in physics after he arrived at Cambridge University. Here the narrative, written for a British readership, may strike American readers as disjointed. Some of the discussions seem calculated to correct misapprehensions or lay to rest controversies about Hawking and his work that occupied the British press and popular culture but are not so well known in the United States. At bottom, the authors are clearly determined to humanize Hawking and show that he is not simply the morality-play figure of a wheelchair-bound genius who transforms all misfortunes into success and fame. End notes; index. 304 pp.

Wigner, Eugene P., and Andrew Szanton. *The Recollections of Eugene P. Wigner.* New York: Plenum Press, 1992.
Memoirs. Wigner (1902-1995) recorded his recollections during interviews with Szanton, a science writer. Szanton edited them and added some published materials. The voice that emerges is modest but sophisticated—an Old World intellectual talking about a great scientific revolution, which he helped bring about. Born in Hungary, Wigner studied or worked with many of the greatest physicists and mathematicians of the twentieth century, including Albert Einstein, David Hilbert, John von Neumann, Edward Teller, and Enrico Fermi. He contributed fundamental ideas to quantum mechanics and became a Nobel laureate for it; he directed an

important laboratory during the Manhattan Project and participated in the development of atomic energy. The book is anecdotal and light fare scientifically; Wigner's opinion of his contemporaries is congenial and sometimes surprising, as when he praises Teller for his sense of humor, something which does not emerge so clearly from the accounts of others. Photos; bibliography; index. 335 pp.

Williams, L. Pearce. *Michael Faraday*, New York: Basic Books, 1965.
A masterfully written scientific biography. Faraday, known as the great experimenter, rose from humble origins to become one of the supreme scientists of the nineteenth century. He uncovered fundamental properties of electricity and magnetism during half a century of research at the Royal Institution in London. Williams' thesis is that the traits of independent thinking, unswerving devotion to his work, painstaking efforts with his experiments, and caution in announcing his findings derive from a lively and warm family and, especially, his upbringing in a strict Christian sect, the Sandemanians. Williams excels at explaining how Faraday pieced together the nature of electromagnetism, often quoting Faraday's publications at length. Photos, drawings, and diagrams; notes at chapter ends; index. 531 pp.

Yukawa, Hideki. *"Tabibito" (The Traveler)*. Singapore: World Scientific Publishing, 1982.
Autobiography. Born Hideki Ogawa, Hideki Yukawa (1907-1981) was one of the architects of the Standard Model of particle physics and known above all for his theoretical prediction that a special particle, the meson, mediates the weak nuclear interaction. The meson theory won him the 1949 Nobel Prize for Physics. He was a national figure in Japan even before that. He taught at Princeton University and Columbia University, directed the Research Institute for Fundamental Physics at Kyoto University, and boldly worked for nuclear disarmament. In this gentle, evocative book he describes his childhood and education and leads the reader to the time when he made his intuitive leap about mesons in 1934. An appendix reprints the paper in which he first presented the

ideas. The text is filled with unassuming wisdom; even readers with almost no knowledge of physics will see why Yukawa is revered in Japan and by fellow physicists. Photos. 218 pp.

Chapter 9

Related Fields

BIOGRAPHIES AND AUTOBIOGRAPHIES

Anderson, H. Allen. *The Chief: Ernest Thompson Seton and the Changing West*. College Station: Texas A&M Press, 1986.
Biography. To Anderson, the widely read artist and nature writer Ernest Thompson Seton (1860-1946) was half Natty Bumpo and half eastern dude. This strange mixture helped Seton to study wild animals and observe the changes wrought to western Canada and the United States by the increase in human population. While Anderson discusses Seton's contributions to science and nature art and literature, he focuses primarily on Seton's personal life and role as a social commentator. For readers who grew up reading Seton's tales, this book is an enlightening look at him. Photos and drawings; bibliography; end notes; index. 363 pp.

Bergengren, Erik. *Alfred Nobel*. London: Thomas Nelson and Sons, 1962.
Biography. Given the forewords by Dag Hammarskjöld and Winston Churchill, one might expect this biography to be an idealization of Alfred Nobel (1833-1896) as the great benefactor of civilization through his invention, dynamite, and the prizes he inaugurated in his will. It is, largely. Bergengren acknowledges that Nobel's hatred of war, belief that dynamite would make it impossible, and drive for wealth evidence a complicated personality, but he warns readers from taking Nobel as a romantic figure. Instead, he was a "versatile, gifted, and self-taught technician, modest, steadfast, firm-

principled, and courageous." This view persists throughout the book, which provides clear explanations of Nobel's inventions and business interests. An appendix discusses the administration and rules for the prizes. Photos; bibliography; index. 222 pp.

Bush, Vannevar. *Pieces of the Action*. New York: William Morrow, 1970.
Memoirs. Advisor to Presidents Franklin Roosevelt and Harry Truman, pioneer computer designer, director or chairman of leading institutions, Vannevar Bush was a mover in scientific circles through the middle of the twentieth century. Here he ruminates on his experiences, which include a key role in creating the Manhattan Project, and discusses the famous politicians and scientists he worked with. He arranged the book by subjects, not chronologically. The first chapter surveys his career in science, and then he discourses on organizations, stumbling blocks to science, the difference between amateurs and professionals, inventions and inventors, energy, teaching, and leadership. Index. 366 pp.

Cutright, Paul Russell. *Theodore Roosevelt the Naturalist*. New York: Harper and Brothers, 1956.
Biography. Cutright examines the fruits of Theodore Roosevelt's love of nature. Beginning with Roosevelt's self-education in natural history as a boy and his studies in biology at Harvard University, Cutright follows his friendship with naturalists; expeditions in the West, Africa, and Amazonia; and his work as president to launch the national park system. Written for a general audience, the book succeeds in showing a facet of Roosevelt's character often obscured by the bully hero of San Juan Hill legend. Photos; bibliography; end notes; index. 297 pp.

De Kruif, Paul. *The Sweeping Wind*. New York: Harcourt, Brace and World, 1962.
Memoirs. This book is more a confession than an autobiography by Paul De Kruif, author of the popular science classics *The Microbe Hunters* (see Chapter 5) and *Men Against Death* (see Chapter 7). De Kruif trained as a microbiologist and had a

promising career in research at the Rockefeller Institute, but he fell in love, and in order to earn the living necessary to support his beloved, he quit scientific research and took up science writing. Much of this book involves the love story, his guilt over abandoning his first wife, and his dealings with other writers, including Sinclair Lewis, whom he helped write *Arrowsmith*. However, de Kruif also discusses the operation of the Rockefeller Institute, which he criticizes for being too elitist, and speaks frankly, sometimes satirically, of fellow scientists and science administrators, such as Simon Flexner. He also describes biochemical research in the early twentieth century. 246 pp.

Earl of Birkenhead. *The Professor and the Prime Minister*. Boston, Mass.: Houghton Mifflin, 1962.
Biography. Frederick Lindemann (1886-1957) trained as a chemist and became a professor at the Clarendon Laboratory at Oxford University, but he is far better known as the maven of scientific politics in Great Britain, serving as science advisor to Winston Churchill throughout World War II. He is responsible for helping many eminent European scientists escape fascist persecution and finding jobs for them in England and elsewhere. The author devotes most of this book to Lindemann (later Viscount Cherwell) as a science administrator, including his participation in England's atomic power program. Photos; index. 400 pp.

Egerton, Judy. *George Stubbs, Anatomist and Animal Painter*. London: Tate Gallery, 1976.
Biography. George Stubbs (1724-1806) earned his living painting the beloved animals, mainly horses, of eighteenth-century gentry. Possessing an unusually keen eye and analytical mind, his paintings and etchings showed anatomical details unlike never before. He also drew the musculature and skeletons of various animals, earning a place in the history of anatomy. This pamphlet was written for an exhibition of Stubbs' works at the Tate Gallery, and it reveals the intricacy and beauty of his style in its illustrations, which are unfortunately few. It also contains essays about Stubbs as an anatomist and scientist, extracts of a memoir of him by the

contemporary artist Ozias Humphry, and a catalog of his art. Photos of his drawings. 64 pp.

Evlanoff, Michael, and Marjorie Fluor. *Alfred Nobel, the Loneliest Millionaire*. Los Angeles, Calif.: Ward Richie Press, 1969.
Biography. Alfred Nobel, according to the authors, was a mass of contradictions, a lonely man moving in a world of indignity and ruthlessness, yet he initially hoped his invention, dynamite, would prevent war. He soon saw he was wrong. The prizes he founded in his will attempt to redeem his error. The authors describe Nobel's scientific achievements, his "pathetically sad life," and his large family. They also list the winners in the various Nobel Prize categories. Photos; index. 336 pp.

Fant, Kenneth. *Alfred Nobel*. New York: Arcade Publishing, 1993.
Biography. A Swedish author and actor, Fant draws upon voluminous letters that Alfred Nobel wrote and received. In fact, Fant frequently excerpts them, which lends intimacy to the text. Although Fant recounts the invention of dynamite and the last testament that inaugurated the Nobel Prizes, he does not dwell on scientific principles. Much of the story involves Nobel's relations with his father and many brothers, his mistress, and his political and business affairs. Photos; index. 342 pp.

Feyerabend, Paul. *Killing Time*. Chicago, Ill.: University of Chicago Press, 1995.
Autobiography. Trained as a physicist and astronomer, Paul Feyerabend (1924-1994) became a leading philosopher of science. This book tells the haphazard events that led him to that career. A harsh childhood in Austria, crippling wounds fighting for Germany at the Russian front during World War II, studies in science and voice, and a passion for opera—none of these may strike the reader as preparation for his famous critique of the scientific method. In fact, it was a chance encounter with the philosopher Karl Popper that set him on his influential course. Told with sharp wit and disregard of conventions, this book contains refreshing views of European science and scientists—such as Niels Bohr and Felix

Ehrenhaft—as well as reflections on science and rationality. Photos; index. 192 pp.

Jaki, Stanley L. *Uneasy Genius: The Life and Work of Pierre Duhem.* The Hague, The Netherlands: Martinus Nijhoff Publishers, 1984.
Biography. Pierre Duhem (1861-1916) conducted research in thermodynamics, for which he gained an international reputation. But Jaki is interested in Duhem's larger place in intellectual history and spends as much time discussing Duhem as a historian and philosopher of science. Jaki carefully traces Duhem's development from childhood to his teaching career at universities in Rennes and Bordeaux before taking up his science, historical research, and philosophy in separate chapters. Jaki writes primarily for historians of science and culture. Photos; bibliography; footnotes; indexes. 472 pp.

Martin, R. N. D. *Pierre Duhem: Philosophy and History in the World of a Believing Physicist.* La Salle, Ill.: Open Court, 1991.
Biography. After a brief introductory chapter outlining the life and career of Pierre Duhem, best known to scientists for his study of thermodynamics, Martin discusses Duhem's systematic study of science history during the last part of his career. Duhem brought to light the sophistication of medieval science especially, as Martin shows. Martin also takes up Duhem's defense of Catholicism and the critical reception of his ideas. For readers interested in intellectual history. Bibliography; end notes; index. 274 pp.

Morrison-Low, A. D., and J. R. R. Christie, eds. *'"Martyr of Science": Sir David Brewster 1781-1868.* Edinburgh, Scotland: Royal Scottish Museum, 1984.
Biography. A neglected figure, David Brewster exerted great influence on science and scientific organizations during the nineteenth century. Some of the ten essays in this volume examine the broad outlines of his life and his scientific interests, which included the nature of light and scientific instrumentation. Most, however, examine his scientific journalism, reform of universities, and participation in such scientific organizations. Photos; bibliography; end notes. 138 pp.

Name Index

Boldface page numbers indicate biographies or autobiographies.

Subject Index

About the Author

Roger Smith is a freelance writer specializing in science and language. He attended Reed College, the University of Nevada-Reno, the University of Copenhagen, and Stanford University, where he graduated with a doctorate in English in 1987. As well as serving in the Marine Corps, he has worked as a magazine editor and a legislative reporter for the Associated Press and has taught writing, literature, and linguistics at the University of Nevada, Fairleigh Dickinson University, Willamette University, and Linfield College. His first book, *Popular Physics and Astronomy: An Annotated Bibliography*, was published by Scarecrow Press and Salem Press in 1996.